Construction Hazardous Materials Compliance Guide

Mold Detection, Abatement, and Inspection Procedures

R. Dodge Woodson

ELSEVIER

AMSTERDAM • BOSTON • HEIDELBERG • LONDON
NEW YORK • OXFORD • PARIS • SAN DIEGO
SAN FRANCISCO • SINGAPORE • SYDNEY • TOKYO
Butterworth-Heinemann is an imprint of Elsevier

Butterworth-Heinemann is an imprint of Elsevier
225 Wyman Street, Waltham, MA 02451, USA
The Boulevard, Langford Lane, Kidlington, Oxford, OX5 1 GB, UK

Notices
Knowledge and best practice in this field are constantly changing. As new research
and experience broaden our understanding, changes in research methods, professional
practices, or medical treatment may become necessary.

Practitioners and researchers must always rely on their own experience and knowledge
in evaluating and using any information, methods, compounds, or experiments described
herein. In using such information or methods they should be mindful of their own
safety and the safety of others, including parties for whom they have a professional
responsibility.

To the fullest extent of the law, neither the Publisher nor the authors, contributors,
or editors, assume any liability for any injury and/or damage to persons or property as a
matter of products liability, negligence or otherwise, or from any use or operation of any
methods, products, instructions, or ideas contained in the material herein.

Library of Congress Cataloging-in-Publication Data
Application submitted.

British Library Cataloguing-in-Publication Data
A catalogue record for this book is available from the British Library.

ISBN: 978-0-12-415840-5

For information on all Butterworth–Heinemann publications
visit our Web site at *http://store.elsevier.com*

Printed in the United States
12   13   14   15   16   10 9 8 7 6 5 4 3 2 1

This book is dedicated to Afton and Adam for always being there for me.

Contents

Introduction

Contractors often encounter mold. It is common in all types of buildings. Not all contractors know how to deal with toxic mold. They should be aware that mold can cause serious health hazards for some people. Whether you are called to solve a mold problem or on a job for another reason and discover a mold problem, you need to be informed on the subject.

The laws, regulations, and rules governing working with mold are set forth by both federal and state agencies. The requirements are far less structured than they are for other types of toxic conditions, but they do exist. Any failure to comply with the requirements detailed by governing authorities can result in stiff fines and potential lawsuits.

R. Dodge Woodson's name is synonymous with professional reference books. Woodson is the author of many bestselling books and has written dozens of valuable texts for the building trades. In this book he provides invaluable insight and guidance in working with mold. There is a broad mix of information available here at your fingertips, ranging from Woodson's field experience to rules, regulations, and laws on both state and federal levels.

If you are someone who has any chance of encountering toxic mold in your work, you need this invaluable guide to keep you informed and safe. Look at the Table of Contents. Thumb through the pages. You will see quickly that this is a comprehensive guide for all types of contractors. Do not let yourself be taken off guard. Read this book and prepare yourself for dealing with mold hazards in future job sites.

▶ ACKNOWLEDGMENTS

I want to thank the U.S. Occupational Safety and Health Administration, the State of Texas, the Centers for Disease Control and Prevention, the U.S. Environmental Protection Agency, and the states of Connecticut and Minnesota for information used in the preparation of this book.

► ABOUT THE AUTHOR

R. Dodge Woodson was a career contractor with more than 30 years of experience. He has been a master plumber, builder, and remodeling contractor since 1979. The Woodson name is synonymous with professional reference books. R. Dodge Woodson has written numerous bestselling books over the years.

Basics about Mold

This chapter provides basic information about mold, sources of mold, and building-related illnesses. Brief discussions are included on building design considerations for healthy indoor air, as well as building evaluation and sampling for mold. However, detailed information about indoor air-quality diagnostic studies (e.g., normal vs. abnormal levels) and the design and execution of exposure sampling strategies is not included as this information is beyond the scope of this book.

For approaches to remediation of moldy areas and the appropriate response based on the degree of the contamination, the reader should consult OSHA's Safety and Health Information Bulletin (SHIB 03-10-10) "A Brief Guide to Mold in the Workplace." Additional information on mold is available through OSHA's "Molds and Fungi" safety and health topics webpage at *www.osha.gov/SLTC/molds/index.html*, which contains a collection of hyperlinks to various sources of information regarding mold.

The Occupational Safety and Health Association's guidance document is not a standard or regulation, and it creates no new legal obligations. It is advisory in nature, informational in content, and is intended to provide relevant information to building owners, managers, and occupants regarding mold prevention and remediation in buildings. Contractors and other professionals (e.g., environmental consultants and health or safety professionals) who respond to mold and moisture situations in buildings, as well as members of the general public, also may want to refer to these guidelines.

Employers are required to comply with hazard-specific safety and health standards as issued and enforced by either the federal Occupational Safety and Health Administration (OSHA), or an OSHA-approved State Plan. In addition, Section 5(a)(1) of the *Occupational Safety and Health Act*, the General Duty Clause, requires employers to provide their employees with a workplace free from recognized hazards likely to cause death or serious physical harm. Employers can be cited for violating the General Duty Clause if there is such a recognized hazard and they do not take reasonable steps to prevent or abate the hazard. However, failure to implement these guidelines is not, in itself, a violation of the General Duty Clause. Citations can only be based on standards, regulations, and the General Duty Clause.

In 1994, OSHA published in the *Federal Register* a comprehensive proposed rule on indoor air quality (IAQ), which addressed adverse health effects attributable to environmental tobacco smoke (ETS) and other indoor pollutants, including bioaerosols (59 FR 15968). During the IAQ rulemaking, the agency received comments and scientific and technical information on indoor mold exposures associated with building-related illnesses (BRIs). While the indoor air quality proposed rule was withdrawn in its entirety in December 2001 (66 FR 64946), the agency retained the voluminous docket (consisting of approximately 120,000 documents), which contains valuable information on ETS and other indoor pollutants, such as chemicals, bacteria, and molds.

In preparing the guide OSHA reviewed the IAQ docket and recent scientific literature pertaining to mold exposures. As a result of this review, it is clear that the entrance of water (i.e., incursion) into buildings that are damaged, poorly designed, or improperly maintained, is the main source of mold-associated building-related illness. Consequently, the focus of the review was directed toward preventive measures to reduce potential environments for mold growth at the source.

The purpose of the guide is to help owners, managers, and occupants understand and prevent building-related illnesses associated with mold problems in offices and other indoor workplaces. It is not the intent of the guide to address the special considerations of building designers, developers, and similar building professionals; however, they may find certain general information helpful. In addition, health care professionals, maintenance workers, custodians, and others who have a role in the prevention and correction (i.e., remediation) of mold problems in buildings may derive benefit from the information and recommendations outlined here.

► MOLD

Molds are the most common forms of fungi found on the earth. Fungi are classified as neither plants nor animals, and include yeasts, mildews, puffballs, and mushrooms. Most molds reproduce through the formation of spores, tiny microscopic cells that float through the indoor and outdoor air on a continual basis. We are all exposed to mold spores in the air we breathe on a daily basis, both indoors and outdoors. See Box 1.1. When mold spores land on a moist surface indoors, they may begin to grow and digest the surface. Left unchecked, molds can eventually destroy the surfaces they grow on. Molds can be any color.

Molds, their fragments, and metabolic by-products have been associated with adverse health effects. Some diseases are known to be caused by specific molds. However, in many occupational settings health conditions suspected to be mold-related cannot be linked to a specific mold as the only possible cause. In a well-known case an initial finding that *Stachybotrys chartarum* (also known as *S. atra*) was linked to acute pulmonary hemorrhage/hemosiderosis in infants living in a water-damaged environment in Cleveland, Ohio, was subsequently disproved. See Box 1.2.

Box 1.1 What Are Molds?

Molds are fungi that can be found both indoors and outdoors. No one knows how many species of fungi exist but estimates range from tens of thousands to perhaps 300,000 or more. Molds grow best in warm, damp, and humid conditions, and spread and reproduce by making spores. Mold spores can survive harsh environmental conditions, such as dry conditions, that do not support normal mold growth.

Box 1.2 How Common Is Mold, Including *Stachybotrys chartarum*?

Molds are very common in buildings and homes and will grow anywhere indoors where there is moisture. The most common indoor molds are *Cladosporium*, *Penicillium*, *Aspergillus*, and *Alternaria*. We do not have precise information about how often *Stachybotrys chartarum* is found in buildings and homes. While it is less common than other mold species, it is not rare.

► WHERE MOLDS ARE FOUND

Molds are found almost everywhere in our environment, both outdoors and indoors. Their spores float continually in the air we breathe. Molds can grow on just about any substance, as long as moisture and oxygen are available. The following are some common indoor molds:

- *Cladosporium*
- *Penicillium*
- *Alternaria*
- *Aspergillus*

Mold growth may occur when excessive moisture accumulates in buildings or on building materials including carpet, ceiling tile, insulation, paper, wallboard, wood, surfaces behind wallpaper, or in heating, ventilation, and air conditioning (HVAC) systems.

► THE CAUSES OF MOLDS IN BUILDINGS

It is impossible to eliminate all molds and mold spores in the indoor environment. However, moisture control is the most important strategy for reducing indoor mold growth. Common sources of moisture in buildings include plumbing, roof, and window leaks; flooding; condensation on cold surfaces (e.g., pipe sweating); poorly maintained drain pans; and wet foundations due to landscaping or gutters that direct water into or under the building. Water vapor from unvented or poorly vented kitchens, showers, combustion appliances, or steam pipes can also create conditions that promote mold growth.

Mold can grow wherever there is dampness. Damp or wet building materials and furnishings should be cleaned and dried within 24 to 48 hours to prevent the growth of mold. See Box 1.3.

► MOLD CONCERNS

Building owners and managers, among others, make numerous decisions about design, operation, and maintenance throughout the life cycles of their buildings. Structural damage to buildings from mold growth is one concern for building owners and managers. If sources of moisture are not controlled, mold, which is always present to some degree, can spread and cause damage to building materials, finishes, and furnishings. Additionally, some molds can cause structural damage to wood.

Box 1.3   Mold Prevention Tips

- Keep the humidity level in your home between 40% and 60%. Use an air conditioner or a dehumidifier during humid months and in damp spaces, such as basements.
- Be sure your home has enough ventilation. Use exhaust fans that vent outside your home in the kitchen and bathroom. Make sure your clothes dryer vents outside your home.
- Fix any leaks in your home's roof, walls, or plumbing so mold does not have moisture to grow.
- Clean up and dry out your home thoroughly and quickly (within 24–48 hours) after flooding.
- Add mold inhibitors to paints before painting.
- Clean bathrooms with mold-killing products.
- Remove or replace carpets and upholstery that have been soaked and cannot be dried promptly. Consider not using carpet in rooms or areas such as bathrooms or basements that may have a lot of moisture.

Structural damage, however, is not the only concern. Large amounts of mold growth in buildings can create odors and may trigger health effects, such as allergic reactions, in some individuals. Illnesses that are associated with mold exposures in buildings are described in this chapter. The results show that, in general, relationships between poor indoor air quality due to the presence of mold and building-related illnesses (BRIs) are unclear. This stems, in part, from the lack of standardized and meaningful methods by which to measure mold exposures and their effects on occupants. However, widespread symptoms related to a building can lead to environmental investigation, mitigation activities, relocation of occupants, and loss of tenants or property value. Problems that follow an onset of health complaints associated with buildings may impact employers located in buildings and sometimes the building owners who may have to bear high costs to resolve the underlying issues.

Information on indoor mold exposures is constantly changing. As new and critical information develops, building professionals and occupants who access the information will be able to incorporate the information into successful resolution of any existing building mold problems.

Building-Related Illnesses

The term *building-related illness* (BRI) is used to describe illnesses characterized by objective clinical findings related to specific exposures in the indoor environment. BRIs are diagnosed by evaluation of signs and symptoms by physicians or other licensed health care professionals. Mold-related BRIs result from mold contamination that has occurred in buildings under specific conditions. All BRIs are preventable by eliminating and controlling the conditions that can lead to the harmful exposures.

How Sick Building Syndrome Differs from BRI
Terms such as *sick building syndrome* (SBS) have been used to describe situations in which building occupants experience a variety of symptoms that, unlike BRIs, appear to be linked to time spent in a building, but no specific illness or cause can be identified. Symptoms often disappear after occupants leave the building.

BRIs Linked to Mold Exposure
The health effects of concern from exposure to mold contamination in an indoor environment can be common allergic BRIs such as allergic rhinitis, allergic asthma, and hypersensitivity pneumonitis (also known as extrinsic allergic alveolitis), and infections such as histoplasmosis and cryptococcosis. Mycotoxins can also produce toxin-mediated adverse health effects. The following discussions of selected mold-related BRIs are not intended to be comprehensive, that is, the descriptions do not include diagnostic tests or medical treatments. Rather, the discussions are informational and focused on common BRIs.

Mold-Related Illnesses
Most people experience no health effects from exposure to the molds present in indoor or outdoor air. However, some individuals with underlying health conditions may be more sensitive to molds. For example, individuals who have other allergies or existing respiratory conditions such as asthma, sinusitis, or other lung diseases may be more easily affected. Similarly, persons who have a weakened immune system tend to be more sensitive to molds.

A person's immune system can be weakened if the individual has conditions such as pregnancy, diabetes, autoimmune disease, leukemia, or AIDS; if the individual is recovering from recent surgery, receiving chemotherapy, or in long-term treatment with steroids; or if the individual is the recipient of a recent organ or bone-marrow transplant. In addition, infants, children, and the elderly have been shown to be more susceptible to health problems attributable to molds.

The most common health effects associated with mold exposure include allergic reactions similar to common pollen or animal allergies. Symptoms include sneezing, runny nose, eye irritation, coughing, congestion, aggravation of asthma, and skin rash. These symptoms are also common reactions to other agents that cause allergies, and it is not always possible to single out the specific cause. More severe health reactions, such as hypersensitivity pneumonitis, can occur in susceptible individuals. The three types of adverse health effects in humans caused by mold are allergy, infection, and toxin-mediated conditions.

▶ PREVENTIVE MAINTENANCE

The key to mold prevention is moisture control. The most important initial step in prevention is a visual inspection. Regular checks of the building envelope and drainage systems should be made to assure that they are in working order. Identify and, to the extent possible, eliminate sources of dampness, high humidity, and moisture to prevent mold growth. Wet or damp spots and wet, nonmoldy materials should be cleaned and dried as soon as possible (preferably within 24 to 48 hours of discovery).

Moisture due to condensation may be prevented by increasing the surface temperature of the material where condensation is occurring, or by reducing the moisture level in the air (humidity). To increase the material's surface temperature, insulate it from the colder area or increase air circulation of warmer air. To reduce the moisture level in the air, repair leaks, increase ventilation (if outside air is cold and dry), or dehumidify (if outside air is warm and humid). Indoor relative humidity should be maintained below 70% (25–60%, if possible).

All buildings should be checked routinely for water leaks, problem seals around doors and windows, and visible mold in moist or damp parts of the building. Any conditions that could be causes of mold growth should be corrected to prevent future mold problems.

Other prevention tips include venting moisture-generating appliances, such as dryers, to the outside where possible; venting kitchens (cooking areas) and bathrooms according to local code requirements; providing adequate drainage around buildings and sloping the ground away from the building foundations; and pinpointing areas where leaks have occurred, identifying the causes, and taking preventive action to ensure that they do not reoccur.

Preventing Mold and Bacterial Growth in Ventilation Systems

Ventilation systems should be checked regularly, particularly for damp filters and overall cleanliness. A preventive maintenance plan should be put into place for each major component of the building's ventilation system. Contact your equipment supplier or manufacturer for recommended maintenance schedules and operations and maintenance manuals. Components that are exposed to water (e.g., drainage pans, coils, cooling towers, and humidifiers) require scrupulous maintenance to prevent microbial growth and the entry of undesired microorganisms or chemicals into the indoor air stream.

Cleaning Air Ducts

Air duct cleaning generally refers to the cleaning of various heating and cooling system components of forced air systems. The components of these systems may become contaminated with mold if moisture is present within the system, resulting in the potential release of mold spores throughout the build-ing. All components of the system must be cleaned. Failure to clean a component of a contaminated system can result in recontamination of the entire system. Water-damaged or contaminated porous materials in the ductwork or other air-handling system components should be removed and replaced. Ventilation system filters should be checked regularly to ensure that they are seated properly. Filters should be replaced on a routine schedule.

▶ PROTECTING OCCUPANTS DURING RENOVATIONS OR REMODELING

The best strategy is to isolate the building area(s) undergoing renovations from occupied areas. Isolating the renovated area(s) usually means erecting barriers made of either plywood or poly-ethylene sheeting. Supply and return ducts should be covered in the area under renovation to prevent the spread of odors and construction dust. Air-handling units serving areas under renovation should also be turned off if they serve only the area being renovated. Air-handling units that are being serviced as part of the renovation should be locked out while they are being serviced. Ensure that the renovated area is under negative or neutral pressure in relation to adjacent occupied space. Evaluate work areas for potential harm to workers and relocate occupants as needed; prevent contamination from spreading to occupied areas.

When undertaking renovations that break the integrity of the building envelope, such as roofing work, regular checks should be made for water intrusions at the breaks in the envelope. Water damage and standing water should be cleaned up immediately.

▶ BUILDING EVALUATION

If you suspect that your building has mold problems, you should look for and eliminate the source of moisture problems in the building. As stated, moisture problems can have many sources, including uncontrolled humidity, roof leaks, and landscaping or gutters that direct water into or under the building. Unvented combustion appliances and standing water following a flood are other sources.

In addition, you should remove all visible mold. Visible mold on external surfaces, especially on the walls of a building, may be an indicator of more severe contamination beneath the surface. However, mold removal without also the correction of the underlying water/moisture problem would not be effective since the mold would just grow back. If a greater problem is suspected, or a moisture problem has resulted in extensive fungal growth, an environmental investigation with emphasis on physical inspection is recommended. An experienced professional should be consulted to evaluate the situation and recommend or supervise the proper corrective action.

▶ SELECTING QUALIFIED PROFESSIONALS

Occupational safety and health professionals are typically able to evaluate a building for mold, whereas occupational health care professionals are qualified to assess and treat illnesses and injuries.

OSHA Publication 3160, *The Occupational Health Professional's Services and Qualifications: Questions and Answers*, provides a thorough discussion of the roles of occupational health physicians, occupational health nurses, industrial hygienists, industrial engineers, safety professionals, and other occupational health professionals. This document is available at *www.osha.gov/Publications/osha3160.pdf*.

Occupational physicians must have completed additional training in occupational medicine beyond the qualifications necessary for medical doctor or doctor of osteopathy licensure. Physicians may be certified in the field after meeting rigorous qualifying standards and successfully completing the examination in occupational medicine given by the American Board of Preventive Medicine. The American College of Occupational and Environmental Medicine (*www.acoem.org*) maintains a directory of some 6000 physicians and other health professionals specializing in occupational and environmental medicine. Your general practice physician may be able to refer you to a specialist from this list.

The American Industrial Hygiene Association (*www.aiha.org*) and the American Society of Safety Engineers (*www.asse.org*) maintain lists of consulting firms for occupational safety and health. Certified Industrial Hygienists (CIHs) and Certified Associate Industrial Hygienists (CAIHs) must have at least a bachelor's degree with a concentration in the sciences, have five years professional experience, and pass a rigorous certification examination. The American Board of Industrial Hygiene (*www.abih.org*) maintains a complete listing of all CIHs and CAIHs in good standing with the organization.

Certified Safety Professionals (CSPs) also must meet academic requirements, have at least four years of experience, and pass a rigorous certification examination. A roster of CSPs is maintained by the Board of Certified Safety Professionals. Both CIHs and CSPs are required to attend continuing education courses to stay current in their field of practice.

▶ SAMPLING FOR MOLD

Where visible mold is present, cleanup can proceed on the basis of the visual inspection. Sampling for molds and other bioaerosols is not usually necessary. There are currently no governmental or professional recommendations for airborne concentrations of mold, mold spores, mycotoxins, and other bioaerosols with which to compare any sampling results. However, sampling for mold may be considered in the following situations:

- When the medical diagnosis is consistent with mold-associated illness
- To delimit the outer boundaries of severely contaminated areas before and during a mold cleanup project
- After a cleanup, to show that the types and concentrations of mold in the area are similar to background levels

Sampling for mold, mold spores, mycotoxins, and other bioaerosols are not part of a routine building evaluation.

Mold Sampling Strategies

Sampling and analysis of mold are complex and can become expensive. There is a lack of standard procedures for sampling and analysis. Sampling should be undertaken only after careful delineation of the sampling goals. For assistance with mold sampling, consult an experienced health and safety professional. Health and safety professionals, working closely with an accredited environmental microbiology laboratory, can determine and document the details concerning the necessary sampling strategy, including when and where to sample.

Standardized methods, such as ACGIH, AIHA, NIOSH, and OSHA methods, should be followed where available. Accredited laboratories that participate in the AIHA Environmental Microbiology Proficiency Analytical Testing (EMPAT) Program are listed on the AIHA website at *www.aiha.org*.

Mold Sampling Results that Are in CFU/m^2 and CFU/m^3

Sampling results for viable (living) microorganisms are presented as concentrations, and the units used will vary depending on the sample collection methods. Air sampling results are reported as colony-forming units per cubic meter of air (CFU/m^3). Specialized sampling is reported in terms of the entity collected, that is, if only spores were sampled, the results would be reported as spores/m^3. Bulk samples may be reported as colony-forming units per gram (CFU/g) of dust or material or colony-forming units per milliliter (CFU/ml) of stagnant water or slime. Wipe sample results are reported as colony-forming units per surface area such as CFU/m^2 or CFU/ft^2. These units represent the culturable portion of mold concentrations only and do not quantify the fragments and by-products of mold that may also exist.

▶ MOLD CONTROL AND REMEDIATION

The purpose of mold remediation is to identify and correct the water or moisture problem, remove or clean all contaminated materials, prevent the spread of contamination to other areas, and protect the health and safety of the building occupants. During any remediation, the health and safety of remediation workers must also be a priority. In every case of microbial contamination, including mold contamination, the underlying cause of the contamination must be rectified or the growth will recur.

These are the basic principles of mold remediation. For more thorough discussions of methods, recommendations, and remediation approaches for various levels of contamination, see OSHA's Safety and Health Information Bulletin entitled, "A Brief Guide to Mold in the Workplace" (SHIB 03-10-10), (1) which is available at: *www.osha.gov/dts/shib/shib101003.html*.

In particular, see discussions concerning:

- Additional measures for cleaning contaminated ductwork
- Biocides vs. antimicrobial agents
- Informing building occupants about mold remediation
- Informing remediation employees about the hazards of mold
- Personal protective equipment (PPE) for remediation employees

▶ WHAT TO DO ABOUT MOLD IN THE WORKPLACE

There are no standards that say how much mold is hazardous to your health. However, there should not be visible mold growth or objectionable moldy odors in your workplace. If you see or smell mold, or if you or others are experiencing mold-related health problems, report the problem to your employer so the problem can be investigated. If you believe that there is a mold hazard, you have the right to file a complaint with Federal OSHA or, in states with OSHA-approved state plans, the state occupational safety and health agency. You can contact your local Area Office of Federal OSHA or state occupational safety and health office or file a complaint online at *www.osha.gov/as/ opa/worker/complain.html.*

Links to the addresses and telephone numbers of the various state occupational safety and health agency offices are available online at *www.osha.gov/fso/osp/index.html.* In addition, assistance with filing complaints, receiving workplace health and safety information, and requesting OSHA publications, among other types of information, are available by calling OSHA's toll-free number at 1-800-321-6742.

▶ GETTING MORE INFORMATION FOR MOLD-RELATED PROBLEMS

The following sources provide links to additional programs and information regarding mold:

Occupational Safety and Health Administration (OSHA); *www. osha.gov*

> Search "indoor air quality" or "molds and fungi" to link to sources of information related to Indoor Air Quality (IAQ) and mold.

Environmental Protection Agency (EPA); *www.epa.gov/iaq/pubs/*

> Indoor Air Quality Information Clearinghouse (IAQ Info), P.O. Box 37133, Washington, DC 20013-7133; 1-800-438-4318; iaqinfo@ aol.com

National Institute for Occupational Safety and Health (NIOSH); *www.cdc.gov/niosh*

> Education and Information Division, Publications Dissemination, 4676 Columbia Parkway, Cincinnati, OH 45226-1988; 1-800-35-NIOSH 1-800-356-4674; pubstaft@cdc.gov

International Facility Management Association; *www.ifma.org*

> 1 E. Greenway Plaza, Suite 1100, Houston, TX 77046-0194; 713-623-4362

American Society of Heating, Refrigerating, and Air Conditioning Engineers (ASHRAE); *www.ashrae.org*

> 1791 Tullie Circle, N.E., Atlanta, GA 30329; 1-800-527-4723

Relevant Publications

Relevant publications used as references for public documents were located through a PubMed (National Library of Medicine) search. PubMed was selected because of its broad scope of publications: Medline (National Library of Medicine), other life sciences journals, and links to sites providing full-text articles. Keywords used to execute the search included mold, epidemiology, BRI, hypersensitivity pneumonitis, farmer's lung, air conditioner or humidifier lung, mycotoxins, and respiratory illness.

The size of the search results was managed, where appropriate, by limiting searches to the English language and human studies. Single word and Boolean logic operators (AND, OR, and NOT) were used to link words and phrases for more precise searches which yielded summary abstracts (from a choice of display formats). Search summaries were quickly reviewed and further refined for useable selections.

Articles were obtained either directly from online sources or through the OSHA Technical Data System and were reviewed both by the Occupational Safety and Health Association staff and contractors. Similar searches of the OSHA IAQ Docket (H-122) were made through the Intra-OSHA Document Management System (DMS). This system contains the entire IAQ docket in a format that allows full-text selection of documents and returns a list of exhibit numbers that contain the term(s) of interest, which can then be viewed for usefulness and retrieved as desired.

Articles were chosen for their subject matter and relevance to workplace building-related illness and were reviewed by OSHA and its contractors. Not all papers reviewed were used.

▶ ADVERSE HEALTH EFFECTS

Molds can cause three types of adverse health effects in humans: allergy, infection, and toxin-mediated conditions. These adverse health effects are discussed in more detail next.

Allergic Rhinitis

Allergic rhinitis has several signs and symptoms. The symptoms of allergic rhinitis include sneezing; itchy eyes, nose, and throat; a stuffy or runny nose; sore throat; cough; watery eyes; headache; and fatigue. These symptoms may be worse in indoor environments and may peak in hot and humid seasons. Common physical findings include red or bloodshot eyes, a runny nose, watery eyes, and thickened nasal mucous membranes. Multiple airborne allergens other than molds (e.g., pollens, animal dander, and dust mites) may be involved. However, when a person is also sensitized to mold, an indoor environment contaminated with mold spores may aggravate the signs and symptoms of allergic rhinitis.

Allergic Asthma

Allergic asthma is a chronic lung disease that is caused by breathing in substances known to be allergens (sensitizers). Asthma occurs when the airways become inflamed and the surrounding muscles tighten and contract. Asthma causes breathing difficulties by making it harder for air to flow in and out of the lungs. Symptoms of asthma include tightness of the chest, shortness of breath, coughing, and wheezing. Although a complete patient history and physical exam are the most important methods employed to diagnose asthma, chest X-rays, pulmonary function testing, and immunologic studies are occasionally also used.

Exposure to mold clearly plays a role in asthma. Molds produce a large variety of compounds that are potentially allergenic (e.g., proteins). Potency varies among these allergens. The epidemiological evidence from well-conducted studies documents the sensitizing potential of mold allergens. Also, well-documented evidence indicates that fungal sensitization is related to asthma.

Hypersensitivity Pneumonitis

Hypersensitivity pneumonitis (HP), or extrinsic allergic alveolitis, refers to a group of allergic lung diseases caused by the inhalation of antigens contained in a wide variety of organic dusts. Antigens are substances that provoke an allergic reaction (i.e., an immune response with the formation of antibodies) when introduced into the body. HP occurs when very small antigenic particles (5 micrometers or less in diameter and invisible to the naked eye) penetrate into the deepest areas of the lungs (the alveoli) and cause inflammation.

HP may occur as an acute, subacute, or chronic condition. Symptoms and signs of the acute form may occur 4 to 6 hours after significant antigen exposure and include chest tightness, difficulty breathing, fever, chills, cough, and muscle aches. If exposure to the antigen is repeated frequently, scars can form on the lungs, resulting in a disabling condition known as fibrosis. Symptoms and signs associated with the subacute and chronic forms of the disease include cough, shortness of breath, fatigue, and weight loss.

A variety of organic dust sources may cause HP. The causative agents may include bacteria, fungi, animal proteins, insects, and chemicals. The numerous clinical syndromes are generally given names reflecting the circumstances or sources of exposure. For example, farmer's lung results from exposures to moldy hay, straw, grain, and compost. Pigeon breeder's or bird-fancier's disease/lung results from exposures to avian protein in bird dander, feathers, and droppings.

Even though HP is not commonly reported as a BRI, sources of concern for building owners and managers include molds, bacteria, and other organisms growing in heating, ventilating, and air conditioning systems, humidifiers, and water damaged buildings. To a lesser extent, HP may be associated with exposure to birds or bird droppings during building cleaning and maintenance activities.

How HP Differs from Asthma

HP differs from asthma in several ways including pathology, diagnosis, and treatment. For example, HP differs from asthma in the location of the inflammation. Asthma is characterized by inflammation of the larger airways close to the mouth and nose. HP is characterized by inflammation of the smallest airways (bronchioles) and the air sacs (alveoli).

Ventilation Pneumonitis

Ventilation pneumonitis (also called air-conditioner lung, humidifier lung, and humidifier fever) is HP that is due to fungal and microbial growth in ventilation and air conditioning systems. Onset can occur when maintenance work is performed on these systems. Diagnosis is based on a combination of characteristic symptoms, chest X-ray findings, pulmonary function abnormalities, and sometimes immunologic study findings. Inspection of the heating, ventilation, and air-conditioning (HVAC) systems and confirmation of the diagnoses are more useful than sampling for mold.

Systemic Fungal Infections

Histoplasmosis and cryptococcosis are examples of infections caused by fungi (*Histoplasma capsulatum* and *Cryptococcus neoformans*, respectively). The main source of exposure to both organisms is debris around bird roosts and soil contaminated with bird and bat droppings. Concern about health risks may be warranted in situations where there is a significant accumulation of bird or bat droppings near ventilation systems and in attics. Both infections are primarily seen in immune-compromised individuals such as those with AIDS, but can also occur in normal healthy individuals.

Toxin-Mediated Conditions

Mycotoxins are metabolic by-products produced by some molds that can cause toxic reactions in humans or animals. Some mycotoxins are concentrated on or within mold spores. Mycotoxins may be hazardous through ingestion, inhalation, or skin contact. The most well-known and studied mycotoxin is aflatoxin. Although not included within the scope of this document, aflatoxin is one of the most potent liver carcinogens known and has been found on contaminated peanuts, grains, and other human and animal foodstuffs.

A wide variety of molds, even some of those most commonly found molds that are generally considered harmless, are capable of producing mycotoxins. Some molds can produce several mycotoxins. Mycotoxin production varies widely depending on the species and the growth conditions, such as availability of nutrients, the suitability of the surface on which growth can take place, environmental factors (e.g., relative humidity, temperature, light, oxygen, and carbon dioxide), the season, maturity of the fungal colony, and competition with other microorganisms.

The presence of mycotoxin-producing mold in a building does not necessarily mean that mycotoxins are present or that building occupants have been exposed to mycotoxins. Mycotoxins are generally not volatile (i.e., do not become airborne easily), and according to recent studies, mycotoxins have not been shown to cause health problems for occupants at concentrations usually seen in residential or commercial buildings.

Adverse health effects that may be due in part to mycotoxins have been reported to occur among agricultural workers. However, these effects are due to the inhalation of very high levels of molds (e.g., in silage and spoiled grain products) that are orders of magnitude greater than the typical exposures that might be seen when mold is found growing indoors.

Toxic Mold

The phrase "toxic mold" has been used by journalists and many others to refer to molds that have been implicated in severe health effects in humans. Although not a scientific term, it is typically used in the press to refer to those molds capable of producing mycotoxins and incorrectly implies that these molds are more dangerous than others. In fact, all molds under the right conditions have the potential to cause allergic reactions, infections, and toxin-mediated conditions. Individuals who believe that they are suffering from mold-related symptoms should seek medical attention. Although there is no singular medical specialty that addresses the indoor air environment, an occupational medicine physician has specialty training addressing diseases associated with the work environment.

▶ OSHA ASSISTANCE

OSHA can provide extensive help through a variety of programs, including technical assistance about effective safety and health programs, state plans, workplace consultations, voluntary protection programs, strategic partnerships, training and education, and more. An overall commitment to workplace safety and health can add value to your business, to your workplace, and to your life.

Safety and Health Program Management Guidelines

Effective management of worker safety and health protection is a decisive factor in reducing the extent and severity of work-related injuries and illnesses and their related costs. To assist employers and employees in developing effective safety and health programs, OSHA published recommended Safety and Health Program Management Guidelines (54 Federal Register 3904-3916, January 26, 1989). These voluntary guidelines apply to all places of employment covered by OSHA.

The guidelines identify four general elements that are critical to the development of a successful safety and health management program:

- Management leadership and employee involvement
- Work analysis
- Hazard prevention and control
- Safety and health training

The guidelines recommend specific actions under each element to achieve an effective safety and health program. They can be viewed on OSHA's website at *www.osha.gov* under the heading Federal Registers.

State Programs

The *Occupational Safety and Health Act of 1970* (OSH Act) encourages states to develop and operate their own job safety and health plans. States with plans approved by OSHA under

Section 18(b) of the OSH Act must adopt standards and enforce requirements that are at least as effective as federal requirements. Currently, there are 26 states that have plans: 22 administer plans covering both private and public (state and local government) employees; the others—Connecticut, New Jersey, New York, and the Virgin Islands—have plans for public-sector employees only.

Consultation Services

Consultation assistance is available on request to employers who want help in establishing and maintaining a safe and healthful workplace. It is largely funded by OSHA and provided as a service at no cost to the employer. Primarily developed for smaller employers with more hazardous operations, the consultation service is delivered by state governments employing professional safety and health consultants. Comprehensive assistance includes an appraisal of all mechanical systems, work practices and occupational safety and health hazards of the workplace, and all aspects of the employer's present job safety and health program.

The program is separate from OSHA's inspection efforts. No penalties are proposed or citations issued for hazards identified by the consultant. The service is confidential. For more information concerning consultation assistance, see the OSHA website at *www.osha.gov/dcsp/smallbusiness/consult.html*.

Voluntary Protection Programs

Voluntary Protection Programs (VPPs) and onsite consultation services, when coupled with an effective enforcement program, expand worker protection to help meet the goals of the Occupational Safety and Health Act. The three levels of VPP—Star, Merit, and Star Demonstration—are designed to recognize outstanding achievement by companies that have successfully incorporated comprehensive safety and health programs into their total management system. The VPPs motivate others to achieve excellent safety and health results in the same outstanding way as they establish a cooperative relationship among employers, employees, and OSHA. For additional information on VPPs and how to apply, visit OSHA's website at: *www.osha.gov/dcsp/vpp/index.html* or contact your nearest OSHA area or regional office.

Strategic Partnership Program

OSHA's Strategic Partnership Program, the newest of OSHA's cooperative programs, helps encourage, assist, and recognize the efforts of partners to eliminate serious workplace hazards and achieve a high level of worker safety and health. Whereas OSHA's Consultation Program and VPP entail one-on-one relationships between OSHA and individual worksites, most strategic partnerships seek to have a broader impact by building cooperative relationships with groups of employers and employees. These partnerships are voluntary, cooperative relationships between OSHA, employers, employee representatives, and others (e.g., labor unions, trade and professional associations, universities, and other government agencies). For more information on this and other cooperative programs, contact your nearest OSHA office, or visit OSHA's website at *www.osha.gov*.

Alliance Programs

The Alliance Program enables organizations committed to workplace safety and health to collaborate with OSHA to prevent injuries and illnesses in the workplace. OSHA and the Alliance participants work together to reach out to, educate, and lead the nation's employers and their employees in improving and advancing workplace safety and health.

Groups that can form an alliance with OSHA include employers, labor unions, trade or professional groups, educational institutions, and government agencies. In some cases, organizations may be building on existing relationships with OSHA through other cooperative programs.

There are few formal program requirements for Alliances and the agreements do not include an enforcement component. However, OSHA and the participating organizations must define, implement, and meet a set of short- and long-term goals that fall into three categories: training and education; outreach and communication; and promoting the national dialogue on workplace safety and health.

Training and Education

OSHA's area offices offer various information services, such as compliance assistance, publications, audiovisual aids, technical advice, and speakers for special engagements. See Box 1.4. OSHA's Training Institute in Arlington Heights, IL, provides basic

---

Box 1.4   Ten Things You Should Know about Mold

1. Potential health effects and symptoms associated with mold exposures include allergic reactions, asthma, and other respiratory complaints.
2. There is no practical way to eliminate all mold and mold spores in the indoor environment; the way to control indoor mold growth is to control moisture.
3. If mold is a problem in your home or school, you must clean up the mold and eliminate sources of moisture.
4. Fix the source of the water problem or leak to prevent mold growth.
5. Reduce indoor humidity (to 30–60%) to decrease mold growth by: venting bathrooms, dryers, and other moisture-generating sources to the outside; using air conditioners and dehumidifiers; increasing ventilation; and using exhaust fans whenever cooking, dishwashing, and cleaning.
6. Clean and dry any damp or wet building materials and furnishings within 24 to 48 hours to prevent mold growth.
7. Clean mold off hard surfaces with water and detergent, and dry completely. Absorbent materials, such as ceiling tiles, that are moldy, may need to be replaced.
8. Prevent condensation: Reduce the potential for condensation on cold surfaces (i.e., windows, piping, exterior walls, roof, or floors) by adding insulation.
9. In areas where there is a perpetual moisture problem, do not install carpeting (i.e., by drinking fountains, by classroom sinks, or on concrete floors with leaks or frequent condensation).
10. Molds can be found almost anywhere; they can grow on virtually any substance, providing moisture is present. There are molds that can grow on wood, paper, carpet, and foods.

---

and advanced courses in safety and health for federal and state compliance officers; consultants; federal agency personnel; and private sector employers, employees, and their representatives.

The OSHA Training Institute has established OSHA Training Institute Education Centers to address the increased demand for its courses from the private sector and from other federal agencies (see OSHA's website at: *www.osha.gov*). These centers are nonprofit colleges, universities, and other organizations that have been selected after a competition for participation in the program.

OSHA also provides funds to nonprofit organizations, through grants, to conduct workplace training and education in subjects where OSHA believes there is a lack of workplace training. Grants are awarded annually. Grant recipients are expected to contribute 20% of the total grant cost.

For more information on grants, training, and education, contact OSHA Training Institute, OSHA Directorate of Training and Education, 2020 South Arlington Heights Road, Arlington Heights, IL 60005-4102; phone 847-297-4810; fax 847-297-4874. For further information on any OSHA program, contact the nearest area or regional office.

Information Available Electronically

OSHA has a variety of materials and tools available on its website at *www.osha.gov*. These include e-tools such as Expert Advisors, Electronic Compliance Assistance Tools (e-cats), Technical Links; regulations, directives, and publications; and videos and other information for employers and employees. OSHA's software programs and compliance assistance tools walk you through challenging safety and health issues and common problems to find the best solutions for your workplace. A wide variety of OSHA materials, including standards, interpretations, directives, and more, can be obtained from the OSHA website.

OSHA Publications

OSHA has an extensive publications program. For a listing of free or sales items, visit OSHA's website at *www.osha.gov* or contact the OSHA Publications Office, U.S. Department of Labor, 200 Constitution Avenue, NW, N-3101, Washington, DC 20210; phone 202-693-1888; fax 202-693-2498.

Contacting OSHA

To report an emergency, file a complaint, or seek OSHA advice, assistance, or products, call 800-321-OSHA or contact your nearest OSHA regional or area office. You can also file a complaint online and obtain more information on OSHA federal and state programs by visiting OSHA's website at *www.osha.gov*. For further information on any OSHA program, contact your nearest OSHA area or regional office. Refer to *www.osha.gov* or call 1-800-321-OSHA for contact information for OSHA area offices, OSHA-approved state plans, and OSHA consultation projects.

Molds in the Environment

2

Molds live in the soil, on plants, and on dead or decaying matter. Outdoors, molds play a key role in the breakdown of leaves, wood, and other plant debris. Molds belong to the kingdom Fungi, and unlike plants, they lack chlorophyll and must survive by digesting plant materials, using plant and other organic materials for food. Without molds, our environment would be overwhelmed with large amounts of dead plant matter.

Molds produce tiny spores to reproduce, just as some plants produce seeds. These mold spores can be found in both indoor and outdoor air, and settled on indoor and outdoor surfaces. When mold spores land on a damp spot, they may begin growing and digesting whatever they are growing on to survive. Since molds gradually destroy the things they grow on, you can prevent damage to building materials and furnishings and save money by eliminating mold growth.

Moisture control is the key to mold control. Molds need both food and water to survive; since molds can digest most things, water is the factor that limits mold growth. Molds will often grow in damp or wet areas indoors. Common sites for indoor mold growth include bathroom tile, basement walls, areas around windows where moisture condenses, and near leaky water fountains or sinks. Common sources or causes of water or moisture problems include roof leaks, deferred maintenance, condensation associated with high humidity or cold spots in the building, localized flooding due to plumbing failures or heavy rains, slow leaks in plumbing fixtures, and malfunction or poor design of humidification systems. Uncontrolled humidity can also be a source of moisture leading to mold growth, particularly in hot, humid climates. See Box 2.1.

▶ HEALTH EFFECTS AND SYMPTOMS ASSOCIATED WITH MOLD EXPOSURE

When moisture problems occur and mold grows as a result, building occupants may begin to report odors and a variety of health problems, such as headaches, breathing difficulties,

> ### Box 2.1 Where Are Molds Found?
>
> Molds are found in virtually every environment and can be detected, both indoors and outdoors, year round. Mold growth is encouraged by warm and humid conditions.
> - Outdoors various types can be found in shady, damp areas or in places where leaves or other vegetation is decomposing.
> - Indoors molds can be found where humidity levels are high, such as basements or showers.

skin irritation, allergic reactions, and aggravation of asthma symptoms; all of these symptoms could potentially be associated with mold exposure.

All molds have the potential to cause adverse health effects. Molds produce allergens, irritants, and, in some cases, toxins that may cause reactions in humans. The types and severity of symptoms depend, in part, on the types of mold present, the extent of an individual's exposure, the ages of the individuals, and their existing sensitivities or allergies.

Inhalation Exposure to Molds and Mycotoxins

Specific reactions to mold growth can include the following:

- **Allergic Reactions:** Inhaling or touching mold or mold spores may cause allergic reactions in sensitive individuals. Allergic reactions to mold are common, and these reactions can be immediate or delayed. Allergic responses include hay fever-type symptoms, such as sneezing, runny nose, red eyes, and skin rash (dermatitis). Mold spores and fragments can produce allergic reactions in sensitive individuals regardless of whether the mold is dead or alive. Repeated or single exposure to mold or mold spores may cause previously nonsensitive individuals to become sensitive. Repeated exposure has the potential to increase sensitivity.
- **Asthma:** Molds can trigger asthma attacks in persons who are allergic (sensitized) to molds. The irritants produced by molds may also worsen asthma in nonallergic (nonsensitized) people.
- **Hypersensitivity Pneumonitis:** Hypersensitivity pneumonitis may develop following either short-term (acute) or long-term (chronic) exposure to molds. The disease resembles bacterial pneumonia and is uncommon.

- **Irritant Effects:** Mold exposure can cause irritation of the eyes, skin, nose, throat, and lungs, and can sometimes create a burning sensation in these areas.
- **Opportunistic Infections:** People with weakened immune systems (i.e., those who are immune-compromised or immune-suppressed) may be more vulnerable to infections as a result of mold (as well as more vulnerable than healthy persons to mold toxins). *Aspergillus fumigatus*, for example, has been known to infect the lungs of immune-compromised individuals. These individuals inhale the mold spores which then start growing in their lungs. *Trichoderma* has also been known to infect immune-compromised children. Healthy individuals are usually not vulnerable to opportunistic infections from airborne mold exposure. However, molds can cause common skin diseases, such as athlete's foot, as well as other infections such as yeast infections.

▶ TOXIC MOLDS

Some molds, such as *Aspergillus versicolor* and *Stachybotrys atra (chartarum)*, are known to produce potent toxins under certain circumstances. Although some mycotoxins are well known to affect humans and have been shown to be responsible for human health effects, for many mycotoxins, little information is available, and in some cases research is ongoing. For example, some strains of *Stachybotrys atra* can produce one or more potent toxins. In addition, preliminary reports from an investigation of an outbreak of pulmonary hemorrhage in infants suggested an association between pulmonary hemorrhage and exposure to *Stachybotrys chartarum*. Review of the evidence of this association at CDC resulted in a published clarification stating that such an association was not established. Research on the possible causes of pulmonary hemorrhage in infants continues. Consult the Centers for Disease Control and Prevention (CDC) for more information on pulmonary hemorrhage in infants.

Molds can produce toxic substances called mycotoxins. Some mycotoxins cling to the surface of mold spores; others may be found within spores. More than 200 mycotoxins have been identified from common molds, and others remain to be identified. Some of the molds that are known to produce mycotoxins are commonly found in moisture-damaged buildings. Exposure pathways for mycotoxins can include inhalation, ingestion, or skin contact. Although some mycotoxins are well known to affect humans and have been shown to be responsible for human health effects, for many mycotoxins, little information is available.

Aflatoxin B_1 is perhaps the most well-known and studied mycotoxin. It can be produced by the molds *Aspergillus flavus* and *Aspergillus parasiticus* and is one of the most potent carcinogens known. Ingestion of aflatoxin B_1 can cause liver cancer. There is also some evidence that inhalation of aflatoxin B_1 can cause lung cancer. Aflatoxin B_1 has been found on contaminated grains, peanuts, and other human and animal foodstuffs. However, *Aspergillus flavus* and *Aspergillus parasiticus* are *not* commonly found on building materials or in indoor environments.

Much of the information about the human health effects of inhalation exposure to mycotoxins comes from studies done in the workplace and some case studies or case reports. Many symptoms and human health effects attributed to inhalation of mycotoxins have been reported including: mucous membrane irritation, skin rash, nausea, immune system suppression, acute or chronic liver damage, acute or chronic central nervous system damage, endocrine effects, and cancer. More studies are needed to get a clear picture of the health effects related to most mycotoxins. However, it is clearly prudent to avoid exposure to molds and mycotoxins.

Some molds can produce several toxins, and some produce mycotoxins only under certain environmental conditions. The presence of mold in a building does not necessarily mean that mycotoxins are present or that they are present in large quantities.

Note: Information on ingestion exposure, for both humans and animals, is abundant. A wide range of health effects has been reported following ingestion of moldy foods including liver damage, nervous system damage, and immunological effects.

Microbial Volatile Organic Compounds

Some compounds produced by molds are volatile and are released directly into the air. These are known as microbial volatile organic compounds (MVOCs). Because these compounds often have strong and/or unpleasant odors, they can be the source of odors associated with molds. Exposure to MVOCs from molds has been linked to symptoms such as headaches, nasal irritation, dizziness, fatigue, and nausea. Research on MVOCs is still in the early phase.

Glucans or Fungal Cell Wall Components

Glucans (also known as β-(1->)-D-Glucans) are small pieces of the cell walls of molds which may cause inflammatory lung and airway reactions. These glucans can affect the immune system when inhaled. Exposure to very high levels of glucans or dust mixtures including glucans may cause a flu-like illness known as Organic Dust Toxic Syndrome (ODTS). This illness has been primarily noted in agricultural and manufacturing settings.

Spores

Mold spores are microscopic (2–10 μm), and they are naturally present in both indoor and outdoor air. Molds reproduce by means of spores. Some molds have spores that are easily disturbed and waft into the air and settle repeatedly with each disturbance. Other molds have sticky spores that will cling to surfaces and are dislodged by brushing against them or by other direct contact. Spores may remain and are able to grow for years after they are produced. In addition, whether or not the spores are alive, the allergens in and on them may remain allergenic for years.

OSHA Workplace Guidelines for Mold

3

Concern about indoor exposure to mold has increased along with public awareness that exposure to mold can cause a variety of health effects and symptoms, including allergic reactions. This chapter details OSHA's recommendations for the prevention of mold growth and describes measures designed to protect the health of building occupants and workers involved in mold cleanup and prevention. The information is directed primarily at building managers, custodians, and others responsible for building maintenance, but may also be used as a basic reference for those involved in mold remediation.

By reading this safety and health information, individuals with little or no experience with mold remediation may be able to reasonably judge whether mold contamination can be managed in-house or whether outside assistance is required. The advice of a medical professional should always be sought if there are any suspected health issues. This data will help those responsible for building maintenance in the evaluation of remediation plans. Contractors and other professionals (e.g., industrial hygienists or other environmental health and safety professionals) who respond to mold and moisture situations in buildings, as well as members of the general public, may also find these guidelines helpful.

The information in this chapter is intended only as a summary of basic procedures and is not intended, nor should it be used, as a detailed guide to mold remediation. The OSHA guidelines are subject to change as more information regarding mold contamination and remediation becomes available.

▶ MOLD BASICS

Molds are part of the natural environment. Molds are fungi that can be found anywhere—inside or outside any structure—throughout the year. About 1000 species of mold can be found in the United States, with more than 100,000 known species worldwide:

- Outdoors, molds play an important role in nature by breaking down organic matter such as toppled trees, fallen leaves, and dead animals. We would not have food and medicines, such as cheese and penicillin, without mold.
- Indoors, mold growth should be avoided. Problems may arise when mold starts eating away at materials, affecting the look, smell, and possibly, with respect to wood-framed buildings, affecting the structural integrity of the buildings.

Molds can grow on virtually any substance, as long as moisture or water, oxygen, and an organic source are present. Molds reproduce by creating tiny spores (viable seeds) that usually cannot be seen without magnification. Mold spores continually float through the indoor and outdoor air. Molds are usually not a problem unless mold spores land on a damp spot and begin growing. They digest whatever they land on to survive. There are molds that grow on wood, paper, carpet, foods, and insulation, while other molds feast on the everyday dust and dirt that gather in the moist regions of a building.

When excessive moisture or water accumulates indoors, mold growth often occurs, particularly if the moisture problem remains uncorrected. While it is impossible to eliminate all molds and mold spores, controlling moisture can control indoor mold growth.

All molds share the characteristic of being able to grow without sunlight; mold needs only a viable seed (spore), a nutrient source, moisture, and the right temperature to proliferate. This explains why mold infestation is often found in damp, dark, hidden spaces; light and air circulation dry areas out, making them less hospitable for mold.

Molds gradually damage building materials and furnishings. If left unchecked, mold can eventually cause structural damage to a wood-framed building, weakening floors and walls as it feeds on moist wooden structural members. If you suspect that mold has damaged building integrity, consult a structural engineer or other professional with the appropriate expertise. Since mold requires water to grow, it is important to prevent excessive moisture in buildings. Some moisture problems in buildings have been linked to changes in building construction practices since the 1970s, which resulted in tightly sealed buildings with diminished ventilation, contributing to moisture vapor buildup.

Other moisture problems may result from roof leaks, landscaping, or gutters that direct water into or under a building, or unvented combustion appliance. Delayed or insufficient maintenance may contribute to moisture problems in buildings. Improper maintenance and design of building heating, ventilating, and air-conditioning (HVAC) systems, such as insufficient cooling capacity for an air-conditioning system, can result in elevated humidity levels in a building.

▶ HEALTH EFFECTS

Currently, there are no federal standards or recommendations (e.g., OSHA, NIOSH, EPA) for airborne concentrations of mold or mold spores. Scientific research on the relationship between mold exposures and health effects is ongoing. This section provides a brief overview, but does not describe all potential health effects related to mold exposure. For more detailed information, consult a health professional or your state or local health department.

There are many types of mold. Most typical indoor air exposures to mold do not present a risk of adverse health effects. Molds can cause adverse effects by producing allergens (substances that can cause allergic reactions). Potential health concerns are important reasons to prevent mold growth and to remediate existing problem areas.

The onset of allergic reactions to mold can be either immediate or delayed. Allergic responses include hay fever-type symptoms such as runny nose and red eyes. Molds may cause localized skin or mucosal infections but, in general, do not cause systemic infections in humans, except for persons with impaired immunity, AIDS, uncontrolled diabetes, or those taking immune suppressive drugs. An important reference with guidelines for immuno-compromised individuals can be found at the Centers for Disease Control and Prevention (CDC) website.

Molds can also cause asthma attacks in some individuals who are allergic to mold. In addition, exposure to mold can irritate the eyes, skin, nose, and throat in certain individuals. Symptoms other than allergic and irritant types are not commonly reported as a result of inhaling mold in the indoor environment.

Some specific species of mold produce mycotoxins under certain environmental conditions. Potential health effects from mycotoxins are the subject of ongoing scientific research and are beyond the scope of this document. Eating, drinking, and using tobacco products and cosmetics where mold remediation is taking place should be avoided. This will prevent unnecessary contamination of food, beverage, cosmetics, and tobacco products by mold and other harmful substances within the work area.

▶ **PREVENTION**

Moisture control is the key to mold control. When water leaks or spills occur indoors, act promptly. Any initial water infiltration should be stopped and cleaned promptly. A prompt response (within 24–48 hours) and thorough cleanup, drying, and/or removal of water-damaged materials will prevent or limit mold growth. Box 3.1 lists some tips for mold prevention.

Box 3.1 Mold Prevention Tips

- Repair plumbing leaks and leaks in the building structure as soon as possible.
- Look for condensation and wet spots. Fix all of the source(s) of moisture incursion problem(s) as soon as possible.
- Prevent moisture from condensing by increasing surface temperature or reducing the moisture level in the air (humidity). To increase surface temperature, insulate or increase air circulation. To reduce the moisture level in the air, repair leaks, increase ventilation (if outside air is cold and dry), or dehumidify (if outdoor air is warm and humid).
- Keep HVAC drip pans clean, flowing properly, and unobstructed.
- Perform regularly scheduled building/HVAC inspections and maintenance, including filter changes.
- Maintain indoor relative humidity below 70% (25–60%, if possible).
- Vent moisture-generating appliances, such as dryers, to the outside where possible.
- Vent kitchens (cooking areas) and bathrooms according to local code requirements.
- Clean and dry wet or damp spots as soon as possible, but no more than 48 hours after discovery.
- Provide adequate drainage around buildings and slope the ground away from building foundations. Follow all local building codes.
- Pinpoint areas where leaks have occurred, identifying the causes, and taking preventive action to ensure that they do not reoccur.

The following questions may assist in determining whether a mold problem currently exists:

- Are building materials or furnishings visibly damaged by moisture?
- Have building materials been wet for more than 48 hours?
- Are there existing moisture problems in the building?
- Are building occupants reporting musty or moldy odors?
- Are building occupants reporting health problems that they think are related to mold in the indoor environment?
- Has the building been recently remodeled or has the building use changed?
- Has routine maintenance been delayed or the maintenance plan been altered?

Always consider consulting a health professional to address any employee health concerns.

▶ REMEDIATION PLAN

Remediation includes both the identification and correction of the conditions that permit mold growth, as well as the steps to safely and effectively remove mold-damaged materials (see Box 3.2).

Before planning any remediation process, assess the extent of the mold or moisture problem and the type of damaged materials. If you choose to hire outside assistance to do the cleanup, make sure the contractor has experience with mold remediation. Check references and ask the contractor to follow the recommendations in EPA's publication "Mold Remediation in Schools and Commercial Buildings," or other guidelines developed by professional or governmental organizations.

The remediation plan should include steps to permanently correct the water or moisture problem. The plan should cover the use of appropriate personal protective equipment (PPE). It also should include steps to carefully contain and remove moldy building materials in a manner that will prevent further contamination. Remediation plans may vary greatly depending on the size and complexity of the job, and may require revision if circumstances change or new facts are discovered.

If you suspect that the HVAC system is contaminated with mold, or if mold is present near the intake to the system, contact

Box 3.2   Determine Extent of Mold Growth

Sometimes it is simple to determine the scope of a mold problem, but often it is not easy or obvious. Leaky roofs or walls, widespread and repeated condensation, or wicking of water through a concrete slab, for example, may lead to extensive hidden mold growth. Once the scope has been determined, a strategy for cleaning and repair may be developed by the homeowner, building manager, or contractor, as appropriate in the specific case. That will be a work in progress if the work uncovers additional affected parts of the building. The widely cited New York City guidelines suggest categories based on size of visible mold growth on interior surfaces, with procedures designed for each. Although these categories provide a rough starting point, the guidelines also allow flexibility (for example, containment alternatives need not be considered if occupants will not be exposed and the work area will be clean before reentry):

- Level I is a small isolated area, 10 square feet of mold growth or less. Examples include ceiling tiles or small areas on walls.
- Level II is a mid-sized, isolated area, 10 to 30 square feet of mold growth. Examples include individual wallboard panels.
- Level III is a large isolated area, 30 to 100 square feet of mold growth. An example is several wallboard panels.
- Level IV is extensive contamination, greater than 100 contiguous square feet of visible mold growth in an area.
- Level V is for remediation of an HVAC system, and is itself divided into subcategories by size of affected area (up to 10 square feet of visible mold growth, and 10 square feet of visible mold growth or over).

the National Air Duct Cleaners Association (NADCA), or consult EPA's guide "Should You Have the Air Ducts in Your Home Cleaned?" before taking further action. Do not run the HVAC system if you know or suspect that it is contaminated with mold, as it could spread contamination throughout the building. If the water or mold damage was caused by sewage or other contaminated water, consult a professional who has experience cleaning and repairing buildings damaged by contaminated water. For more information on mold-contaminated HVAC systems, see Boxes 3.3 through 3.6.

The remediation manager's highest priority must be to protect the health and safety of the building occupants and remediators. Remediators should avoid exposing themselves and others to mold-laden dusts as they conduct their cleanup activities. Caution should be used to prevent mold and mold spores from being dispersed throughout the air where they can be inhaled by building occupants (see Box 3.7). In some cases, especially those

> ### Box 3.3 Steps before Cleaning and Remediation of HVAC Systems
>
> • If the building is to remain partly occupied (for example, on upper floors not affected by flood waters), isolate the construction areas where HVAC systems will be cleaned and remediated by using temporary walls, plastic sheeting, or other vapor-retarding barriers. Maintain the construction areas under negative pressure (relative to adjacent nonconstruction areas) by using blowers equipped with HEPA filters (high-efficiency particulate air filters) to exhaust the area. To ensure complete isolation from the construction areas, it may be necessary to pressurize the adjacent nonconstruction areas and temporarily relocate the outdoor-air intake for the HVAC system serving the occupied areas.
>
> • Take precautions to protect the health of workers who are cleaning and remediating the HVAC system. Make sure that workers wear at least an N95 NIOSH-approved respirator to protect against airborne microorganisms. Increased levels of respiratory protection (for example, powered, air-purifying respirators equipped with HEPA filters) may be appropriate depending on the level of visible contamination. In addition, when using chlorine bleach or other disinfectants in poorly ventilated environments, it may be necessary to use appropriate chemical cartridges in addition to the particulate filters to protect workers from breathing the chemical vapors.
>
> Employers must implement a complete respiratory protection program that meets the requirements of the OSHA respiratory protection standard (29 Code of Federal Regulations 1910.134). The minimum requirements for a respiratory protection program include a written standard operating procedure for the following: selecting and using respirators; the medical evaluation of workers to determine whether they are physically able to wear the respirator selected for use; training and instructions on respirator use; the cleaning, repair, and storage of respirators; the continued surveillance of work area conditions for worker exposure and stress; and a respirator fit-testing program. For tight-fitting respirators, fit-testing is necessary to help ensure that the respirator fits tightly, reducing the potential for leakage of outside air from around the edge of the mask. In addition, employers must provide workers with appropriate skin, eye, and hearing protection for the safe performance of their jobs.

involving large areas of contamination, the remediation plan may include temporary relocation of some or all of the building occupants.

When deciding if relocating occupants is necessary, consideration should be given to the size and type of mold growth, the type and extent of health effects reported by the occupants, the potential health risks that could be associated with the

Box 3.4 HVAC Cleaning and Remediation

- Remove all flood-contaminated insulation surrounding and within the HVAC system components. Discard these contaminated materials appropriately following applicable federal, state, and local regulations.
- Remove contaminated HVAC filter media and discard appropriately following applicable federal, state, and local regulations.
- After removing any insulation and filters, clean all flood-contaminated HVAC system component surfaces with a HEPA-filtered vacuum cleaner to remove dirt, debris, and microorganisms. Pay special attention to filter racks, drain pans, bends, and horizontal sections of air ducts where debris can collect.
- After removing any insulation or debris, disinfect all HVAC system component surfaces while the HVAC system is not operating. Use a solution of 1 cup of household chlorine bleach in a gallon of water. Do not mix bleach with other cleaning products that contain ammonia.
- Conduct the cleaning and disinfection activities in a clean-to-dirty work progression. Consider the use of auxiliary fans to supply "clean" air to the worker position and carry aerosolized contaminant and disinfectant in the clean-to-dirty direction, away from the worker's breathing zones and toward the point of filtration and exhaust.
- Follow the disinfection procedure with a clean water rinse. Depending on the amount of debris present, it may be necessary to mechanically clean the HVAC system component surfaces with a steam or a high-pressure washer before using the disinfectant. Gasoline powered pressure washers should be used outside or with adequate exhaust ventilation to prevent carbon monoxide hazards.
- *Note*: Remove and discard HVAC system components that are contaminated with flood water, and cannot be effectively cleaned and disinfected. Replace them with new components.
- After cleaning and disinfecting or replacing the HVAC system components, replace the insulation—preferably with an external (i.e., not in the air stream) smooth-surfaced insulation to help prevent debris and microorganisms from collecting in the future.
- Make sure that the HVAC system fan has been removed and serviced (cleaned, disinfected, dried thoroughly, and tested) by a qualified professional before it is placed back into the air-handling unit.
- During the cleaning and remediation process, consider upgrading the HVAC system filtration to the highest efficiency filters practical given the static pressure constraints of the HVAC system fan. This step has been shown to be one of the most cost-effective ways to improve the long-term quality of the indoor environment, since it reduces the amount of airborne dusts and microorganisms.

remediation activity, and the amount of disruption this activity is likely to cause. In addition, before deciding to relocate occupants, one should also evaluate the remediator's ability to contain/minimize possible aerosolization of mold spores given their expertise and the physical parameters of the workspace. When

> **Box 3.5 Key Engineering Controls and Work Practices for HVAC Systems' Mold Remediation**
>
> • Use the recommended work practices and isolation methods; areas <10 square feet may be considered small isolated areas and area >10 square feet should be considered areas with extensive contamination.
> • Shut down the HVAC system before beginning remedial activities; consult with the building maintenance staff, engineer, or HVAC manufacturer to determine the correct procedures for shutting down the system and to identify/locate system components and areas of potential contamination.
> • Consult with the HVAC manufacturer to determine which biocide they recommend for use on their HVAC system and components such as cooling coils and condensation pans.
> • Remove contaminated materials that can support mold growth, such as the paper on the insulation of interior lined ducts and filters.
> • Inspect and clean HVAC system surfaces by removing all dirt, debris, and visible mold. Disinfect mold-impacted surfaces before reusing the system.
> • For extensively contaminated areas (i.e., >10 square feet of contamination), air monitoring with the HVAC system running should be conducted prior to the reoccupancy

possible, remediation activities should be scheduled during off hours when building occupants are less likely to be affected.

Remediators, particularly those with health-related concerns, may wish to check with their physicians or another health-care professional before working on mold remediation or investigating potentially moldy areas. If any individual has health concerns, doubts, or questions before beginning a remediation/cleanup project, he or she should consult a health professional (see Box 3.8).

Readers with a mold contamination situation should consult the more detailed guidance provided by the resources cited at the end of this chapter. Where the contamination is significant, homeowners, building managers, and maintenance personnel should seek qualified professional advice and assistance. To remediate is to fix or correct a problem, or in other words, to apply a remedy. "Remediation" encompasses actions and techniques to correct growth of mold (fungi) in indoor areas. According to Health Canada (the Canadian health department): "*Remediation*" includes both the thorough cleaning of any mold

Box 3.6 Resuming HVAC Operations

- After cleaning and disinfecting or replacing the HVAC system, have a qualified professional thoroughly evaluate its performance and correct it as necessary before the building is occupied again. The HVAC system performance should conform to the recommendations contained in ASHRAE Standard 62.1-2007, Ventilation for Acceptable Indoor Air Quality.
- Before the building is occupied again, operate the HVAC system continuously in a normal manner at a comfortable temperature for 48 to 72 hours. During this period, it may be beneficial to open the HVAC outdoor air dampers to the maximum setting that still allows you to provide the desired indoor air temperatures. If objectionable flood-related odors persist after this "flush out" period, reassess by looking for flood-contaminated areas that were not identified earlier and continue the flush-out process until odors are no longer apparent. Replace the HVAC filters used during the flush-out prior to building occupancy.
- After a building is occupied again, make frequent (e.g., weekly) checks of the HVAC system to ensure that it is operating properly. During these checks, inspect the HVAC system filters and replace them when necessary. Gradually reduce the frequency of the HVAC system checks to monthly or quarterly inspections, depending on the routine operation and maintenance specifications for the HVAC system.
- If no routine operation and maintenance program is in place for the HVAC system, develop and institute such a program. At a minimum, include the following routine procedures: inspection and maintenance of HVAC components, calibration of HVAC system controls, and testing and balancing of the HVAC system.
- After the building is occupied again, maintain the interior temperature and relative humidity to conform with the ranges recommended in ASHRAE Standard 55-2004, Thermal Environmental Conditions for Human Occupancy.

Box 3.7 Key Engineering Controls and Work Practices for Silica, Mold, Nuisance Dust, Dried Mud, or Silt

- Stay upwind of or away from dust-generating activities, and in particular those involving crystalline silica-containing materials such as concrete, brick, tile, drywall, mortar, sand, or stone. When inhaled, the fine crystalline silica particles contained in the dust can become lodged deep in the lung, which can lead to silicosis and other respiratory illnesses.
- Use water spray or mist to suppress dust generation, especially during operations that may create a lot of dust, such as cutting or sawing silica-containing materials, jack hammering, impact drilling, using heavy equipment, and demolishing structures.
- Avoid using compressed air for cleaning surfaces.
- Sample worker exposures to silica during dust-generating activities.
- Limit contact or disturbance of surfaces containing substantial visible mold growth.

> **Box 3.8 Risks of Remediating Mold**
>
> Mold remediation exposes the persons doing the work—whether home-owners, maintenance personnel, or professional remediators—to mold spores, fragments, and toxins via inhalation, skin and mucous membrane contact, and possible ingestion. Where contamination is more than trivial (a small, confined, easily accessible area), it is essential to use appropriate protective measures. Where there is risk of spreading spores or particles to other areas within a building, it is also essential to use appropriate means of isolation, as remediation activities and air movements spread spores and particles. This provides only an overview of the elements of remediation.

growing in the building and the correction of the defect that led to mold growth—excessive humidity, water leaking, or water infiltration from the outside.

▶ MOLD REMEDIATION/CLEANUP METHODS

The purpose of mold remediation is to correct the moisture problem and to remove moldy and contaminated materials to prevent human exposure and further damage to building materials and furnishings. Porous materials that are wet and have mold growing on them may have to be discarded because molds can infiltrate porous substances and grow on or fill in empty spaces or crevices. This mold can be difficult or impossible to remove completely. As a general rule, simply killing the mold, for example, with biocide is not enough. The mold must be removed, since the chemicals and proteins that can cause a reaction in humans are present even in dead mold.

A variety of cleanup methods are available for remediating damage to building materials and furnishings caused by moisture control problems and mold growth. The specific method or group of methods used will depend on the type of material affected. Some methods that may be used include the following.

Wet Vacuums

Wet vacuums are vacuum cleaners designed to collect water. They can be used to remove water from floors, carpets, and hard surfaces where water has accumulated. They should not be used to vacuum porous materials, such as gypsum board. Wet vacuums should be used only on wet materials, as spores may be exhausted into the indoor environment if insufficient liquid is present. The tanks, hoses, and attachments of these vacuums should be thoroughly cleaned and dried after use since mold and mold spores may adhere to equipment surfaces.

Damp Wipe

Mold can generally be removed from nonporous surfaces by wiping or scrubbing with water and detergent. It is important to dry these surfaces quickly and thoroughly to discourage further mold growth. Instructions for cleaning surfaces, as listed on product labels, should always be read and followed.

HEPA Vacuum

High-efficiency particulate air (HEPA) vacuums are recommended for final cleanup of remediation areas after materials have been thoroughly dried and contaminated materials removed. HEPA vacuums are also recommended for cleanup of dust that may have settled on surfaces outside the remediation area. Care must be taken to assure that the filter is properly seated in the vacuum so that all the air passes through the filter. When changing the vacuum filter, remediators should wear respirators, appropriate personal protective clothing, gloves, and eye protection to prevent exposure to any captured mold and other contaminants. The filter and contents of the HEPA vacuum must be disposed of in impermeable bags or containers in such a way as to prevent release of the debris.

Disposal of Damaged Materials

Building materials and furnishings contaminated with mold growth that are not salvageable should be placed in sealed impermeable bags or closed containers while in the remediation area. These materials can usually be discarded as ordinary construction waste. It is important to package mold-contaminated materials in this fashion to minimize the dispersion of mold spores. Large items with heavy mold growth should be covered with polyethylene sheeting and sealed with duct tape before being removed from the remediation area. Some jobs may require the use of dust-tight chutes to move large quantities of debris to a dumpster strategically placed outside a window in the remediation area. See Boxes 3.9 and 3.10 for more on disposal.

Box 3.9 Key Engineering Controls and Work Practices for Mold on Water-Damaged Materials

- Discard all water-damaged materials, materials that are visibly coated with mold that cannot be properly cleaned, such as porous materials (e.g., carpeting, drywall, insulation), and materials that have been wet for more than 48 hours.
- Wrap and seal the items that will be discarded in plastic bags or sheets to reduce the spread of spores. These materials can usually be discarded as ordinary debris.
- Minimize dust disturbance to reduce the spread of fungal spores.
- Do not eat, drink, or smoke in work areas.
- Provide natural or local exhaust ventilation during all cleaning steps.
- Clean hard and nonporous materials using a detergent. After rinsing, if needed, disinfect with an appropriate biocide such as bleach. Don't mix bleach with ammonia-containing products.
- After an area has been cleaned and is completely dry, vacuum the area with a high-efficiency particulate air (HEPA) vacuum. HEPA vacuums are also recommended for cleaning up dust that may have settled on surfaces outside the work area.

Box 3.10 Remove Damaged Porous Materials

Moldy porous materials, such as drywall and carpets, are often removed from the premises and discarded. Mold extends filaments (hyphae) into a porous substrate and therefore may not be removed by surface treatments. Exceptions would be limited to high-value items, such as artwork and Oriental rugs, that might be successfully cleaned professionally by specialists. Treatment may be considered for some gypsum products (e.g., Sheetrock with minor surface mold caused by elevated humidity, or party walls exposed to rain after being framed in). Safe removal requires bagging in airtight plastic bags and removal from the building for conveyance to a sanitary landfill. Moldy materials are not considered "hazardous waste" and may be disposed of with other trash, but they should be securely bagged as a precaution as if they contain moldy compost.

Use of Biocides

The use of a biocide, such as chlorine bleach, is not recommended as a routine practice during mold remediation, although there may be instances where professional judgment may indicate its use (for example, when immuno-compromised individuals are present). In most cases, it is not possible or desirable to sterilize an area, as a background level of mold spores comparable to the level in outside air will persist. However, the spores in the ambient air will not cause further problems if the moisture level in the building has been corrected.

Biocides are toxic to animals and humans, as well as to mold. If you choose to use disinfectants or biocides, always ventilate the area, using outside air if possible, and exhaust the air to the outdoors. When using fans, take care not to extend the zone of contamination by distributing mold spores to a previously unaffected area. Never mix chlorine bleach solution with other cleaning solutions or detergents that contain ammonia because this may produce highly toxic vapors and create a hazard to workers.

Some biocides are considered pesticides, and some states require that only registered pesticide applicators apply these products in schools, commercial buildings, and homes. Make sure anyone applying a biocide is properly licensed where required. Fungicides are commonly applied to outdoor plants, soil, and grains as a powder or spray. Examples of fungicides include hexachlorobenzene, organomercurials, pentachlorophenol, phthalimides, and dithiocarbamates.

Do not use fungicides developed for outdoor use in any indoor application, as they can be extremely toxic to animals and humans in an enclosed environment. When you use biocides as a disinfectant or a pesticide, or as a fungicide, you should use appropriate PPE, including respirators. Always read and follow product label precautions. It is a violation of federal (EPA) law to use a biocide in any manner inconsistent with its label direction.

▶ MOLD REMEDIATION GUIDELINES

This section presents remediation guidelines for building materials that have or are likely to have mold growth. The guidelines are designed to protect the health of cleanup personnel and other workers during remediation. These guidelines are based on the size of the area impacted by mold contamination. Please note that these *are* guidelines; some professionals may prefer to use other remediation methods, and certain circumstances may require different approaches or variations on the approaches described in the following. If possible, remediation activities should be scheduled during off-hours when building occupants are less likely to be affected.

Although the level of personal protection suggested in these guidelines is based on the total surface area contaminated and the potential for remediator or occupant exposure, professional judgment always should play a part in remediation decisions. These remediation guidelines are based on the size of the affected area to make it easier for remediators to select appropriate techniques, not on the basis of research showing that there is a specific method appropriate at a certain number of square feet. The guidelines have been designed to help construct a remediation plan. The remediation manager should rely on professional judgment and experience to adapt the guidelines to particular situations. When in doubt, caution is advised. Consult an experienced mold remediator for more information.

Level I: Small Isolated Areas
The following applies to 10 sq. ft or less areas; for example, ceiling tiles and small areas on walls.

- Remediation can be conducted by the regular building maintenance staff as long as they are trained on proper cleanup methods, personal protection, and potential health hazards. This training can be performed as part of a program to comply with the requirements of the OSHA Hazard Communication Standard (29 CFR 1910.1200).
- Respiratory protection (e.g., N95 disposable respirator) is recommended. Respirators must be used in accordance with the OSHA respiratory protection standard (29 CFR 1910.134). Gloves and eye protection should be worn.
- The work area should be unoccupied. Removing people from spaces adjacent to the work area is not necessary, but is recommended for infants (<12 months old), persons recovering from recent surgery, immune-suppressed people, or people with chronic inflammatory lung diseases (e.g., asthma, hypersensitivity pneumonitis, and severe allergies).
- Containment of the work area is not necessary. Dust suppression methods, such as misting (not soaking) surfaces prior to remediation, are recommended.
- Contaminated materials that cannot be cleaned should be removed from the building in a sealed impermeable plastic bag. These materials may be disposed of as ordinary waste.

- The work area and areas used by remediation workers for egress should be cleaned with a damp cloth or mop and a detergent solution.
- All areas should be left dry and visibly free from contamination and debris.

Level II: Mid-Sized Isolated Areas

The following applies to 10 to 30 square-foot areas; for example, individual wallboard panels.

- Remediation can be conducted by the regular building maintenance staff. Such persons should receive training on proper cleanup methods, personal protection, and potential health hazards. This training can be performed as part of a program that will comply with the requirements of the OSHA Hazard Communication Standard (29 CFR 1910.1200).
- Respiratory protection (e.g., N95 disposable respirator) is recommended. Respirators must be used in accordance with the OSHA respiratory protection standard (29 CFR 1910.134). Gloves and eye protection should be worn.
- The work area should be unoccupied. Removing people from spaces adjacent to the work area is not necessary, but is recommended for infants (<12 months old), persons recovering from recent surgery, immune-suppressed people, or people with chronic inflammatory lung diseases (e.g., asthma, hypersensitivity pneumonitis, and severe allergies).
- Surfaces in the work area that could become contaminated should be covered with a secured plastic sheet(s) before remediation to contain dust and/or debris and prevent further contamination.
- Dust suppression methods, such as misting (not soaking) surfaces prior to remediation, are recommended.
- Contaminated materials that cannot be cleaned should be removed from the building in a sealed impermeable plastic bag. These materials may be disposed of as ordinary waste.
- The work area and areas used by remediation workers for egress should be HEPA vacuumed and cleaned with a damp cloth or mop and a detergent solution.
- All areas should be left dry and visibly free from contamination and debris.

Level III: Large Isolated Areas

The following applies to 30 to 100 sq. ft areas; for example, several wallboard panels. Industrial hygienists or other environmental health and safety professionals with experience performing microbial investigations and/or mold remediation should be consulted prior to remediation activities to provide oversight for the project. These procedures may be implemented depending on the severity of the contamination.

- It is recommended that personnel be trained in the handling of hazardous materials and equipped with respiratory protection (e.g., N95 disposable respirator). Respirators must be used in accordance with the OSHA respiratory protection standard (29 CFR 1910.134). Gloves and eye protection should be worn.
- Surfaces in the work area and areas directly adjacent that could become decontaminated should be covered with a secured plastic sheet(s) before remediation to contain dust/debris and prevent further contamination.
- Seal ventilation ducts/grills in the work area and areas directly adjacent with plastic sheeting.
- The work area and areas directly adjacent should be unoccupied. Removing people from spaces near the area is recommended for infants, persons having undergone recent surgery, immune-suppressed people, or people with chronic inflammatory lung diseases. (e.g., asthma, hypersensitivity pneumonitis, and severe allergies).
- Dust suppression methods, such as misting (not soaking) surfaces prior to mediation, are recommended.
- Contaminated materials that cannot be cleaned should be removed from the building in sealed impermeable plastic bags. These materials may be disposed of as ordinary waste.
- The work area and surrounding areas should be HEPA vacuumed and cleaned with a damp cloth or mop and a detergent solution.
- All areas should be left dry and visibly free from contamination and debris.

Note: If abatement procedures are expected to generate a lot of dust (e.g., abrasive cleaning of contaminated surfaces, demolition of plaster walls) or the visible concentration of the mold is heavy (blanket coverage as opposed to patchy), it is recommended that the remediation procedures for Level IV be followed.

Level IV: Extensive Contamination

The following applies to an area greater than 100 contiguous square feet. Industrial hygienists or other environmental health and safety professionals with experience performing microbial investigations and/or mold remediation should be consulted prior to remediation activities to provide oversight for the project. These procedures may be implemented depending on the severity of the contamination.

- Personnel should be trained in the handling of hazardous materials and equipped with:
  - Full face piece respirators with HEPA cartridges
  - Disposable protective clothing covering entire body including both head and shoes
  - Gloves
- Containment of the affected area should include:
  - Complete isolation of work area from occupied spaces using plastic sheeting sealed with duct tape (including ventilation ducts/grills, fixtures, and other openings)
  - The use of an exhaust fan with a HEPA filter to generate negative pressurization
  - Airlocks and decontamination room
- If contaminant practices effectively prevent mold from migrating from affected areas, it may not be necessary to remove people from surrounding work areas. However, removal is still recommended for infants, persons having undergone recent surgery, immune-suppressed people, or people with chronic inflammatory lung diseases (e.g., asthma, hypersensitivity pneumonitis, and severe allergies).
- Contaminated materials that cannot be cleaned should be removed from the building in sealed impermeable plastic bags. The outside of the bags should be cleaned with a damp cloth and a detergent solution or HEPA vacuumed in the decontamination chamber prior to their transport to uncontaminated areas of the building. These materials may be disposed of as ordinary waste.
- The contained area and decontamination room should be HEPA vacuumed and cleaned with a damp cloth or mopped with a detergent solution and be visibly clean prior to the removal of isolation barriers.

▶ PERSONAL PROTECTIVE EQUIPMENT

Any remediation work that disturbs mold and causes mold spores to become airborne increases the degree of respiratory exposure, so personal protective equipment (PPE) will be needed. Actions that tend to disperse mold include: breaking apart moldy porous materials such as wallboard; destructive invasive procedures to examine or remediate mold growth in a wall cavity; removal of contaminated wallpaper by stripping or peeling; using fans to dry items or ventilate areas (see Boxes 3.11 and 3.12).

The primary function of personal protective equipment is to prevent the inhalation and ingestion of mold and mold spores and to avoid mold contact with the skin or eyes. The following subsections discuss the various types of PPE that may be used during remediation activities.

Skin and Eye Protection

Gloves protect the skin from contact with mold, as well as from potentially irritating cleaning solutions. Long gloves that extend to the middle of the forearm are recommended. Glove material should be selected based on the type of substance/chemical being handled. If you are using a biocide such as chlorine bleach, or

---

Box 3.11 General PPE Guidelines

PPE is recommended for all response/recovery tasks and/or operations:

- *Minimum:* Gloves, N95 respirator, goggles/eye protection
- *Limited:* Gloves, N95 respirator or half-face respirator with HEPA filter, disposable overalls, goggles/eye protection
- *Full:* Gloves, disposable full body clothing, head gear, foot coverings, full-face respirator with HEPA filter

Additional PPE may be needed for a specific hazard, such as:

- Hard hat for overhead impact or electrical hazards
- Eye protection with side shields
- Gloves chosen for job hazards expected (e.g., heavy-duty leather work gloves for handling debris with sharp edges and/or chemical protective gloves appropriate for chemicals potentially contacted)
- ANSI-approved protective footwear
- Respiratory protection as necessary: N95, R95, or P95 filtering face pieces may be used for nuisance dusts (e.g., dried mud, dirt, and silt) and mold (except mold remediation); filters with a charcoal layer may be used for odors

---

█ Box 3.12   Additional Personal Protective Equipment

- N95, R95, or P95 respirators: Either a half-face or a full-face N95, R95, or P95 respirator for areas <100 square feet; for areas >100 square feet, areas where mold is heavy (blanket coverage rather that patchy), or areas where substantial dust is generated during cleaning or debris removal (e.g., abrasives are used to clean surfaces), use a full-face respirator with N100, R100, or P100 filters. Charcoal-impregnated filters may be used for odors. When silica is being used for abrasive blasting, an abrasive-blasting respirator must be used.
- Nonvented goggles are the best kind of eye protection.
- Protective clothing (e.g., disposable coveralls) to prevent cross contamination and skin contact with mold and chemicals. For areas greater than 100 square feet, ensure that protective clothing covers entire body including head and feet.
- Long gloves made of an appropriate material will protect the user from chemicals handled for surface cleaning.

Respirator Protection

Respirator protection for workers in isolated areas of mold contamination (<100 square feet) or small isolated areas of HVAC systems (<10 square feet) where mold is disturbed should be at least as protective as an N95 filtering face piece. For working in areas of extensive contamination (>100 contiguous square feet) or HVAC systems with large areas of contamination (>10 square feet) and significant mold-containing dust, fullface piece respirators with N100, R100, P100 particulate filters (or for powered air-purifying respirators—HEPA filters) are recommended.

---

a strong cleaning solution, you should select gloves made from natural rubber, neoprene, nitrile, polyurethane, or PVC. If you are using a mild detergent or plain water, ordinary household rubber gloves may be used.

To protect your eyes, use properly fitted goggles or a full face piece respirator. Goggles must be designed to prevent the entry of dust and small particles. Safety glasses or goggles with open vent holes are not appropriate in mold remediation.

Respiratory Protection

Respirators protect cleanup workers from inhaling airborne mold, contaminated dust, and other particulates that are released during the remediation process. Either a half mask or full face piece air-purifying respirator can be used. A full face piece respirator provides both respiratory and eye protection. Refer to the earlier section outlining the different levels of remediation to ascertain the type of respiratory protection recommended. Respirators used to provide protection from mold and mold spores must be certified by the National Institute for Occupational Safety and Health (NIOSH). More protective respirators may have to be selected and used if toxic contaminants such as asbestos or lead are encountered during remediation.

As specified by OSHA in 29 CFR 1910.134, individuals who use respirators must be properly trained, have medical clearance, and be properly fit-tested before they begin using a respirator. In addition, use of respirators requires the employer to develop and implement a written respiratory protection program, with worksite-specific procedures and elements.

Protective Clothing

While conducting building inspections and remediation work, individuals may encounter hazardous biological agents as well as chemical and physical hazards. Consequently, appropriate protective clothing (i.e., reusable or disposable) is recommended to minimize cross-contamination between work and clean areas to prevent the transfer and spread of mold and other contaminants to street clothing, and to eliminate skin contact with mold and potential chemical exposures.

Disposable clothing should be discarded immediately after it is used. They should be placed into impermeable bags, and can usually be discarded as ordinary construction waste. Appropriate precautions and protective equipment for biocide applicators should be selected based on the product manufacturer's warnings and recommendations (e.g., goggles or face shield, aprons or other protective clothing, gloves, and respiratory protection).

▶ SAMPLING FOR MOLD

Is it necessary to sample for mold? In most cases, if visible mold growth is present, sampling is unnecessary. Air sampling for mold may not be part of a routine assessment because decisions about appropriate remediation strategies often can be made on the basis of a visual inspection.

Your first step should be to inspect for any evidence of water damage and visible mold growth. Testing for mold is expensive, and there should be a clear reason for doing so. In many cases, it is not economically practical or useful to test for mold growth on surfaces or for airborne spores in the building. In addition, there are no standards for "acceptable" levels of mold in buildings, and the lack of a definitive correlation between exposure levels and health effects makes interpreting the data difficult, if not impossible.

Testing is usually done to compare the levels and types of mold spores found inside the building with those found outside of the building or for comparison with another location in the building. In addition, air sampling may provide tangible evidence supporting a hypothesis that investigators have formulated. For example, air sampling may show a higher concentration of the same species of mold when the HVAC is operating than when it has been turned off. This finding may convince the investigators that the mold is growing within, and being disseminated by, the HVAC system. Conversely, negative results may persuade investigators to abandon this hypothesis and to consider other sources of mold growth or dissemination. If you know you have a mold problem, it is more important to spend time and resources removing the mold and solving the moisture problem that causes the moldy conditions than to undertake extensive testing for the type and quantity of mold.

If you are in doubt about sampling, consult an industrial hygienist or other environmental health or safety professional with experience in microbial investigations to help you decide if sampling for mold is necessary or useful, and to identify persons who can conduct any necessary sampling. Due to the wide difference in individual susceptibility to mold contamination, sampling results may have limited application. However, sampling results can be used as a guide to determine the extent of an infestation and the effectiveness of the cleanup. Their interpretation is best left to the industrial hygienist or other environmental health or safety professional.

Sampling for mold should be conducted by professionals with specific experience in designing mold-sampling protocols, sampling methods for microbial contaminants, and interpretation of results. For additional information on air sampling, refer to the American Conference of Governmental Industrial Hygienists' document "Bioaerosols: Assessment and Control." In addition, sampling and analysis should follow any other methods recommended by either OSHA, NIOSH, EPA, the

American Industrial Hygiene Association, or other recognized professional guidelines. Types of samples can include: air samples, surface samples, bulk samples, and water samples from condensate drain pans or cooling towers.

Microscopic identification of the spores/colonies requires considerable expertise. These services are not routinely available from commercial laboratories. Documented quality control in the laboratories used for analysis of the bulk, surface, and other air samples is necessary. The American Industrial Hygiene Association offers accreditation to microbial laboratories, such as the Environmental Microbiology Laboratory Accreditation Program (EMLAP). Accredited laboratories must participate in quarterly proficiency testing; for example, the Environmental Microbiology Proficiency Analytical Testing Program (EMPAT).

▶ REMEDIATION EQUIPMENT

There are various types of equipment useful in mold assessment and remediation. Some of the more common items are described in the following sections.

Moisture Meters

Moisture meters measure/monitor moisture levels in building materials, and may be helpful for measuring the moisture content in a variety of building materials following water damage. They also can be used to monitor the progress of drying damaged materials. These direct reading devices have a thin probe which is inserted into the material to be tested or pressed directly against the surface of the material. Moisture meters can be used on materials such as carpet, wallboard, wood, brick, and concrete.

Humidity Gauges or Meters

Humidity meters can be used to monitor indoor humidity. Inexpensive (less than $50) models that monitor both temperature and humidity are available.

Humidistat

A humidistat is a control device that can be connected to an HVAC system and adjusted so that if the humidity level rises above a set point, the HVAC system will automatically turn on and reduce the humidity below the established point.

Boroscope

A boroscope is a hand-held tool that allows users to see potential mold problems inside walls, ceiling plenums, crawl spaces, and other tight areas. It consists of a video camera on the end of a flexible "snake." No major drilling or cutting of dry wall is required.

HVAC System Filter

High-quality filters must be used in an HVAC system during remediation because conventional HVAC filters are typically not effective in filtering particles the size of mold spores. Consult an engineer for the appropriate filter efficiency for your specific HVAC system, and consider upgrading your filters if necessary. A filter with a minimum efficiency of 50 to 60% or a rating of MERV 8, as determined by Test Standard 52.2 of the American Society of Heating, Refrigerating and Air-Conditioning Engineers, may be appropriate. Remember to change filters as appropriate, especially following any remediation activities. Remove filters in a manner that minimizes the reentry of mold and other toxic substances into the workplace. Under certain circumstances, it may be necessary to wear appropriate PPE while performing this task.

▶ HOW TO KNOW WHEN YOU HAVE FINISHED REMEDIATION/CLEANUP

Use the following checklist to determine when remediation and cleanup has been completed.

- You must have identified and completely corrected the source of the water or moisture problem.
- Mold removal should be complete. Visible mold, materials that are mold-damaged, and moldy odors should no longer be present.
- Sampling, if conducted, should show that the level and types of mold and mold spores inside the building are similar to those found outside.
- You should revisit the site(s) after remediation, and it should show no signs of moldy or musty odors, water damage, or mold growth.

► CONCLUSION

After correcting water or moisture infiltration, prompt removal of contaminated material and structural repair is the primary response to mold contamination in buildings. In all situations, the underlying cause of water accumulation must be rectified or the mold growth will reoccur. Emphasis should be placed on preventing contamination through proper building and HVAC system maintenance and prompt repair of water damaged areas.

Effective communication with building occupants is an essential component of all large-scale remediation efforts. The building owner, management, and/or employer should notify occupants in the affected area(s) of the presence of mold. Notification should include a description of the remedial measures to be taken and a timetable for completion. Group meetings held before and after remediation with full disclosure of plans and results can be an effective communication mechanism. Individuals with persistent health problems that appear to be related to mold exposure should see their physicians for a referral to practitioners who are trained in occupational/environmental medicine or related specialties and are knowledgeable about these types of exposures.

This concludes our look at OSHA's recommendations for dealing with mold in the workplace.

Government Resources

American Conference of Governmental Industrial Hygienists, 1999. *Bioaerosols: Assessment and Control. www.acgih.org*

The Building Owners and Managers Association International (BOMA): *www.boma.org*

National Apartment Association: *www.naahq.org*

National Institute for Occupational Safety and Health (NIOSH): *www.cdc.gov/niosh*

National Multi-Housing Council: *www.nmhc.org*

New York City Department of Health & Mental Hygiene Bureau of Environmental & Occupational Disease Epidemiology, 2002. *Guidelines on Assessment and Remediation of Fungi in Indoor Environments.*

U.S. Environmental Protection Agency, Office of Air and Radiation, Indoor Environments Division, 2001. *Mold Remediation in Schools and Commercial Buildings.* EPA 402-K-01-001.

Additional Resources

Business owners who are concerned about the cost of professional help can contact the OSHA Consultation Project Office in their state for free consultation service. Priority is given to businesses with fewer than 250 employees at a worksite, with further consideration given to the severity of the worksite problem. The Consultation Program can help the employer evaluate and prevent hazardous conditions in the workplace that can cause injuries and illnesses, including mold problems.

Air Conditioning Contractors of America (ACCA)
  Information on indoor comfort products and services.
Allergy & Asthma Network Mothers of Asthmatics (AANMA)
  Information on allergies and asthma.
American Academy of Allergy Asthma & Immunology (AAAAI)
  Physician referral directory, information on allergies and asthma.
American College of Occupational and Environmental Medicine (ACOEM)
  Referrals to physicians who have experience with environmental exposures.
American Conference of Governmental Industrial Hygienists, Inc. (ACGIH)
  Occupational and environmental health and safety information.
American Industrial Hygiene Association (AIHA)
  Information on industrial hygiene and indoor air quality issues including mold hazards and legal issues.
American Lung Association (ALA)
  Information on allergies and asthma.
American Society of Heating, Refrigerating and Air Conditioning Engineers, Inc. (ASHRAE)
  Information on engineering issues and indoor air quality.
Association of Occupational and Environmental Clinics (AOEC)
  Referrals to clinics with physicians who have experience with environmental exposures, including exposure to mold; maintains a database of occupational and environmental cases.
Asthma and Allergy Foundation of American (AAFA)
  Information on allergies and asthma.
Carpet and Rug Institute (CRI)
  Carpet maintenance, restoration guidelines for water-damaged carpet, other carpet-related issues.
Centers for Disease Control and Prevention (CDC)
  Information on health-related topics including asthma, molds in the environment, and occupational health. CDC is recognized as the lead federal agency for protecting the health and safety of the American people at home and abroad. It serves as the national focus for developing and applying disease prevention and control, environmental health, and health promotion and education activities.

Floods and Flooding, Federal Emergency Management Agency (FEMA)
Publications on floods, flood proofing, and more.

Indoor Environmental Institute (IEI)
Information on best practices in building remediation.

Institute of Inspection, Cleaning and Restoration Certification (IICRC)
Information on and standards for the inspection, cleaning, and restoration industry.

International Sanitary Supply Association (ISSA)
Education and training on cleaning and maintenance.

National Air Duct Cleaners Association (NADCA)
Duct cleaning information.

National Institute of Allergy and Infectious Diseases (NIAID)
Information on allergies and asthma.

National Institute of Building Sciences (NIBS)
Information on building regulations, science, and technology.

National Institute for Occupational Safety and Health (NIOSH)
Health and safety information with a workplace orientation.

National Jewish Health
Information on allergies and asthma.

National Pesticide Information Center (NPIC)
Information on pesticides/antimicrobial chemicals, including safety and disposal information.

New York City Department of Health and Mental Hygiene, Bureau of Environmental and Occupational Disease Epidemiology
Guidelines on Assessment and Remediation of Fungi in Indoor Environments

Occupational Safety and Health Administration (OSHA)
Information on worker safety and health, compliance assistance, laws and regulations, cooperative programs, state programs, statistics, and newsroom.

Restoration Industry Association (RIA)
Carpet and Upholstery Cleaning Institute, Mechancial Systems Hygiene Institute, National Institute of Disaster Restoration, National Institute Rug Cleaning, Water Loss Institute referrals to professionals.

Sheet Metal and Air Conditioning Contractors' National Association (SMACNA)
Technical information on topics such as air conditioning and air ducts.

University of Minnesota, Department of Environmental Health and Safety
Managing water infiltration into buildings.

U.S. EPA IAQ Information Clearinghouse (IAQINFO)
Indoor air related documents, answers to Indoor Air Quality (IAQ) questions, maintains listing of State IAQ contacts, and regional EPA contacts.

Mold in Homes

4

A variety of molds are part of the natural environment. Outdoors molds play a part in nature by breaking down dead organic matter such as fallen leaves and dead trees, but indoors, mold growth should be avoided. Molds reproduce by means of tiny spores; the spores are invisible to the naked eye and float through outdoor and indoor air. Mold may begin growing indoors when mold spores land on surfaces that are wet. There are many types of mold, and none of them will grow without water or moisture.

Can mold cause health problems? Molds are usually not a problem indoors, unless mold spores land on a wet or damp spot and begin growing. Molds have the potential to cause health problems. They produce allergens (substances that can cause allergic reactions), irritants, and in some cases, potentially toxic substances (mycotoxins). Inhaling or touching mold or mold spores may cause allergic reactions in sensitive individuals. Allergic responses include hay fever-type symptoms, such as sneezing, runny nose, red eyes, and skin rash (dermatitis). Allergic reactions to mold are common. They can be immediate or delayed. Molds can also cause asthma attacks in people with asthma who are allergic to mold. In addition, mold exposure can irritate the eyes, skin, nose, throat, and lungs. Three key points about mold are:

- The key to mold control is moisture control.
- If mold is a problem in your home, you should clean up the mold promptly and fix the water problem.
- It is important to dry water-damaged areas and items within 24 to 48 hours to prevent mold growth.

Symptoms other than the allergic and irritant types are not commonly reported as a result of inhaling mold. Research on mold and health effects is ongoing. This chapter provides a brief overview; it does not describe all potential health effects related to mold exposure. For more detailed information you should consult a health professional. You may also wish to consult your state or local health department.

> ### Box 4.1 Identify and Fix the Causes of Mold
>
> The first step is to stop whatever is causing water intrusion or accumulation, if that can be done before beginning cleanup efforts. Sometimes a wall or other compartment must be opened before the cause can be identified, but often the source of the problem can be identified before building repairs are started. For example, the problem might be as simple as a misaimed sprinkler, a leaking pipe under a sink, or plugged drainage.

► HOW TO GET RID OF MOLD

It is impossible to get rid of all mold and mold spores indoors; some mold spores will be found floating through the air and in house dust. The mold spores will not grow if moisture is not present. Indoor mold growth can and should be prevented or controlled by controlling moisture indoors. If there is mold growth in your home, you must clean up the mold and fix the water problem. If you clean up the mold, but don't fix the water problem, then, most likely, the mold problem will come back (see Box 4.1).

Molds can gradually destroy the things they grow on. You can prevent damage to your home and furnishings, save money, and avoid potential health problems by controlling moisture and eliminating mold growth.

► WHO SHOULD DO THE CLEANUP?

Who should do the cleanup depends on a number of factors. One consideration is the size of the mold problem. If the moldy area is less than about 10 square feet (less than roughly a 3×3 ft patch), in most cases, you can handle the job yourself, following the guidelines below. However, if there has been a lot of water damage, and/or mold growth covers more than 10 square feet, consult the U.S. Environmental Protection Agency (EPA) guide: "Mold Remediation in Schools and Commercial Buildings." Although focused on schools and commercial buildings, this document is applicable to other building types. It is available online at *www.epa.gov/mold*.

If you choose to hire a contractor (or one of many other professional service providers) to do the cleanup, make sure the contractor has experience cleaning up mold. Check references and ask the contractor to follow the recommendations in EPA's

"Mold Remediation in Schools and Commercial Buildings," the guidelines of the American Conference of Governmental Industrial Hygienists (ACGIH), or other guidelines from similar professional or government organizations.

If you suspect that the heating, ventilation, air-conditioning (HVAC) system may be contaminated with mold (it is part of an identified moisture problem, for instance, or there is mold near the intake to the system), consult EPA's guide "Should You Have the Air Ducts in Your Home Cleaned?" before taking further action. Do not run the HVAC system if you know or suspect that it is contaminated with mold as it could spread mold throughout the building. Visit *www.epa.gov/iaq/pubs* to download a copy of the EPA guide. If the water and/or mold damage was caused by sewage or other contaminated water, then call in a professional who has experience cleaning and fixing buildings damaged by contaminated water. If you have health concerns, consult a health professional before starting cleanup.

▶ TIPS AND TECHNIQUES

The tips and techniques presented in this section will help you determine how to clean up your mold problem. Professional cleaners or remediators may use methods not covered in this publication. Please note that mold may cause staining and cosmetic damage. It may not be possible to clean an item so that its original appearance is restored.

- Fix plumbing leaks and other water problems as soon as possible. Dry all items completely.
- Scrub mold off hard surfaces with detergent and water, and dry completely.
- Places that are often or always damp can be hard to maintain completely free of mold. If there's some mold in the shower or elsewhere in the bathroom that seems to reappear, increasing the ventilation (running a fan or opening a window) and cleaning more frequently will usually prevent mold from recurring, or at least keep the mold to a minimum.
- Absorbent or porous materials, such as ceiling tiles and carpet, may have to be thrown away if they become moldy.
- Mold can grow on or fill in the empty spaces and crevices of porous materials, so the mold may be difficult or impossible to remove completely. Avoid exposing yourself or others to mold.

- Do not paint or caulk moldy surfaces. Clean up the mold and dry the surfaces before painting. Paint applied over moldy surfaces is likely to peel.
- If you are unsure about how to clean an item, or if the item is expensive or of sentimental value, you may wish to consult a specialist. Specialists in furniture repair, restoration, painting, art restoration and conservation, carpet and rug cleaning, water damage, and fire or water restoration are commonly listed in phone books. Be sure to ask for and check references. Look for specialists who are affiliated with professional organizations.
- Avoid breathing in mold or mold spores. To limit exposure to airborne mold, you may want to wear an N95 respirator, available at many hardware stores and from companies that advertise on the Internet. Some N95 respirators resemble a paper dust mask with a nozzle on the front, others are made primarily of plastic or rubber and have removable cartridges that prevent most mold spores from entering. To be effective, the respirator or mask must fit properly, so carefully follow the instructions supplied with it. Please note that the Occupational Safety and Health Administration (OSHA) requires that respirators fit properly (fit testing) when used in an occupational setting; consult OSHA for more information (800-321-OSHA or *www.osha.gov*).

How to Know When Remediation or Cleanup Is Finished

You must have completely fixed the water or moisture problem before the cleanup or remediation can be considered finished. You should have completed mold removal. Visible mold and moldy odors should not be present. Please note that mold may cause staining and cosmetic damage. You should have revisited the site(s) shortly after cleanup and it should show no signs of water damage or mold growth. People should have been able to occupy or reoccupy the area without health complaints or physical symptoms. Ultimately, this is a judgment call; there is no easy answer.

Wear Gloves and Goggles

Wear gloves and goggles when working with mold. Long gloves that extend to the middle of the forearm are recommended. When working with water and a mild detergent, ordinary household rubber gloves may be used. If you are using a disinfectant, a biocide (e.g., chlorine bleach), or a strong cleaning solution, you should select gloves made from natural rubber, neoprene, nitrile, polyurethane, or PVC. Avoid touching mold or moldy items with your bare hands. Goggles that do not have ventilation holes are recommended. Avoid getting mold or mold spores in your eyes.

Avoiding Mold

Avoiding mold is easier when you take some precautionary steps. When water leaks or spills occur indoors you should act quickly. If wet or damp materials or areas are dried 24 to 48 hours after a leak or spill happens, in most cases mold will not grow. Clean and repair roof gutters regularly. Make sure the ground slopes away from the building foundation, so that water does not enter or collect around the foundation. Keep air conditioning drip pans clean and the drain lines unobstructed and flowing properly. See Box 4.2 for information about building standards.

Keep indoor humidity low. If possible, keep indoor humidity below 60% (ideally between 30–50%) relative humidity. Relative humidity can be measured with a moisture or humidity meter, a small, inexpensive ($10–50) instrument available at many hardware stores. If you see condensation or moisture collecting on windows, walls, or pipes take steps quickly to dry the wet surface and reduce the moisture/water source. Condensation can be a sign of high humidity. Actions that will help to reduce humidity can include the following:

- Vent appliances that produce moisture, such as clothes dryers, stoves, and kerosene heaters to the outside where possible. (Combustion appliances such as stoves and kerosene heaters produce water vapor and will increase the humidity unless vented to the outside.)

Box 4.2 Building Standards

Indoor water damage and dampness are damaging and costly and can pose health risks. The legislature might wish to examine or request appropriate review of building standards to identify changes that can help to prevent leaks and chronic dampness in buildings. Issues to be examined might include strengthening architectural and engineering emphasis on dampness control and leak prevention, strengthening methods to assure that waterproofing measures are properly installed during construction, and assuring that energy efficiency measures do not result in indoor water accumulation.

- Use air conditioners and/or dehumidifiers when needed.
- Run the bathroom fan or open the window when showering. Use exhaust fans or open windows whenever cooking, running the dishwasher, or dishwashing, etc.
- Increase ventilation or air movement by opening doors and/or windows, when practical. Use fans as needed.
- Cover cold surfaces, such as cold water pipes, with insulation.
- Increase air temperature.

▶ TESTING OR SAMPLING FOR MOLD

Is sampling for mold needed? In most cases, if visible mold growth is present, sampling is unnecessary. Since no EPA or other federal limits have been set for mold or mold spores, sampling cannot be used to check a building's compliance with federal mold standards. Surface sampling may be useful to determine if an area has been adequately cleaned or remediated. Sampling for mold should be conducted by professionals who have specific experience in designing mold sampling protocols, sampling methods, and interpreting results. Sample analysis should follow analytical methods recommended by the American Industrial Hygiene Association (AIHA), the American Conference of Governmental Industrial Hygienists (ACGIH), or other professional organizations.

Suspicion of Hidden Mold

You may suspect hidden mold if a building smells moldy but you cannot see the source, or if you know there has been water damage and residents are reporting health problems. Mold may be hidden in places such as the back side of dry wall, wallpaper, or paneling; the top side of ceiling tiles; the underside of carpets and pads, and so on. Other possible locations of hidden mold include areas inside walls around pipes (with leaking or condensing pipes), the surface of walls behind furniture (where condensation forms), inside ductwork, and in roof materials above ceiling tiles (due to roof leaks or insufficient insulation).

Investigating Hidden Mold Problems

Investigating hidden mold problems may be difficult and will require caution when the investigation involves disturbing potential sites of mold growth. For example, removal of wallpaper can lead to a massive release of spores if there is mold growing on the underside of the paper. If you believe that you may have a hidden mold problem, consider hiring an experienced professional.

▶ CLEANUP AND BIOCIDES

Biocides are substances that can destroy living organisms. The use of a chemical or biocide that kills organisms such as mold (chlorine bleach, for example) is not recommended as a routine practice during mold cleanup. See Box 4.3 for information about cleaners and disinfectants. There may be instances, however, when professional judgment may indicate its use (for example, when immune-compromised individuals are present).

Box 4.3 Cleaners and Disinfectants

For small, minor mold, typically called "mildew," affecting areas, such as shower enclosures, ordinary household cleaning procedures and products are sufficient. In those cases, the mold is growing on a surface layer of grime. Removing that layer of grime also removes the mold. The problem area (assuming the underlying surface is impermeable—tile, for example) may be wiped down with a sponge or cloth and a mild household detergent. A bleach solution (according to package directions) might also be used. (Never mix cleaning products, as some combinations can produce toxic fumes. Bleach and ammonia are specifically not to be used together.)

The sponges and cloths used to clean up mold should be discarded. To take a simple example, a moldy throw-rug on a smooth tile surface can be bagged, removed from the premises, and properly discarded. (It is not considered hazardous waste, but should be bagged—preferably double-bagged—for safety in transit.) Then the tile can be mopped with a mild detergent, rinsed, dried, and ready for use again. If mold has grown on a layer of grime on a smooth surface, the surface can likewise be cleaned. Even after cleaning, stains might remain on surfaces. The stains are not hazardous. Cleaners and disinfectants are not suitable for treating moldy porous surfaces. Spraying such products onto mold both generates product fumes—themselves a potential hazard, especially in insufficiently ventilated areas—and spreads mold spores. The spores are hydrophobic (water resistant), and will be blown into the air by the spray. They are not like ordinary dust or soil particles that can be suppressed by sprayed water.

The NYC Guidelines caution: "The use of gaseous, vapor-phase, or aerosolized biocides for remedial purposes is not recommended. The use of biocides in this manner can pose health concerns for people in occupied spaces of the building and for people returning to the treated space if used improperly. Furthermore, the effectiveness of these treatments is unproven and does not address the possible health concerns from the presence of the remaining nonviable mold."

In most cases, it is not possible or desirable to sterilize an area; a background level of mold spores will remain. These spores will not grow if the moisture problem has been resolved. If you choose to use disinfectants or biocides, always ventilate the area and exhaust the air to the outdoors. Never mix chlorine bleach solution with other cleaning solutions or detergents that contain ammonia because toxic fumes could be produced. Dead mold may still cause allergic reactions in some people, so it is not enough to simply kill the mold, it must also be removed.

▶ BEWARE OF MOLD

Do what you can to avoid mold. Simple steps go a long way in the prevention process. If you get a mold problem, don't ignore it. The longer the mold exists the greater the risk becomes. Act quickly to neutralize the potential health problems associated with mold exposure. Give serious consideration to calling in qualified professionals to deal with your mold problems if they are substantial. Taking this approach has been effective in the past and can be in the future.

General Mold Remediation Methods

5

The first step to consider when conducting mold remediation is communication. There will need to be communication between workers, property owners, and occupants. Special communication strategies may be desirable if you are treating a mold problem in a school. Teachers, parents, and other locally affected groups should be notified of significant issues as soon as they are identified. Consider holding a special meeting to provide parents with an opportunity to learn about the problem and ask questions of school authorities, particularly if it is necessary or advisable to ensure that the school is vacated during remediation. For more information on investigating and remediating molds in schools, refer to EPA's *IAQ Tools for Schools Action Kit* and the asthma companion piece for the *IAQ Tools for Schools* kit, entitled *Managing Asthma in the School Environment*.

Communication with building occupants is essential for mold remediation to be successful. Boxes 5.1 through 5.5 outline recommendations for communications. Some occupants will naturally be concerned about mold growth in their building and the potential health impacts. Occupants' perceptions of the health risk may rise if they perceive that information is being withheld from them. The status of the building investigation and remediation should be openly communicated including information on any known or suspected health risks.

Small remediation efforts will usually not require a formal communication process, but do be sure to take individual concerns seriously and use common sense when deciding whether formal communications are required. Individuals managing medium or large remediation efforts should make sure they understand and address the concerns of building occupants and communicate clearly what has to be done as well as possible health concerns.

Communication approaches include regular memos and/or meetings with occupants (with time allotted for questions and

Box 5.1   Targeting Communications for Homeowners

Mold becomes a serious and widespread problem after floods and heavy, prolonged rain. News coverage of the aftermath of Hurricane Katrina has frequently discussed the prevalence of mold in flooded homes in New Orleans and other parts of the Gulf Coast region. At the time of such events, it is appropriate for public officials to provide information to homeowners and others about mold, the risks it poses, and what to do about it.

Methods of communication include broadcast public service announcements, providing information to print and broadcast media for use in developing news stories, and provision of basic information sheets to be handed out by emergency personnel and by merchants who sell goods that homeowners would typically use in cleaning up and repairing after flooding.

Key information for such communications is encompassed in the California Department of Health Services' flier "Mold in My Home: What Do I Do?" and in the U.S. Environmental Protection Agency's *A Brief Guide to Mold, Moisture, and Your Home*. Those documents include basic information on potential health hazards as well as on basic remediation requirements

Box 5.2   Communicating with Commercial and Government Building Occupants

The California Department of Health Services and California Department of Industrial Relations published a four-page flier titled "Molds in Indoor Workplaces." That flier, available from the Hazard Evaluation System & Information Service (HESIS), concisely summarizes how exposure can happen and what health effects might be associated with it. In April 2002, the Texas Department of Health issued a "Review of Practices for Mold Remediation." The hazard communication section of that document summarizes guidance applicable to workers and building occupants: Information about the potential hazards associated with mold growth in a building and remediation activities should be communicated to both the workers involved in the remediation and the occupants of the building.

NYC DOH recommends training building maintenance staff who will conduct remediation work on the potential health hazards of mold. This training can be conducted as part of the training needed to comply with the OSHA Hazard Communication Standard (29 CFR 1910.1200). Health Canada suggests that building maintenance personnel and maintenance staff be aware of potential problems associated with contaminated indoor air, and USEPA indicates that remediation workers, and particularly those with health-related concerns, might wish to consult with a health-care provider before working on mold remediation or investigating potentially moldy areas. Both USEPA and New York City's DOH recommend

---

Box 5.2 cont'd

communication with building occupants throughout the remediation process. When mold contamination requiring a large-scale response is found, building occupants should be notified of that fact and given a description and timetable of the activities that will take place. The form (e.g., memos, meetings) and extent of communication will depend on the degree of contamination and nature of the remediation work.

The U.S. EPA notes that frequent and open communication maximizes the amount of time available for remediation work by addressing issues and concerns as they arise. At least two levels of communication are appropriate. The first—basic, general information—is of the type in the HESIS flier, and is suitable for distribution at any time for public information purposes. The second, incident-specific information before and during a remediation, as described in the Texas Department of Health document, is appropriate whenever the need arises and is an obligation of building managers, supervisors, and remediation personnel.

Box 5.3 Communicating with Renters

Because of the risks of damage to the dwelling as well as potential health risks from damp indoor environments, both landlords and tenants can benefit from communications about the need to prevent and promptly report water damage. Such communications can be included in rental agreements and posted on the premises.

Box 5.4 Protect Building Occupants

During a mold remediation project, steps need to be taken to protect building occupants from exposure to mold spores and particles and to other contaminants normally stirred up by cleaning, repair, and reconstruction work. Steps in protecting occupants include:

- Inform building occupants of the remediation project, potential risks, and safety precautions.
- Isolate affected areas.
- Remove occupants from adjacent rooms.
- Remove especially susceptible individuals from the site entirely.
- Keep occupants and visitors out of areas being cleaned and repaired.
- Properly dispose of contaminated materials after enclosing them in bags.

Larger projects may require removing occupants from larger areas, such as an entire floor, or even vacating an entire building, especially where the problem encompasses more than just isolated parts of a building.

Box 5.5 Means of Communication

Many methods can communicate information about risks and appropriate responses.

- Brochures and fliers
- Posted signs
- Public service announcements on television or radio
- Press releases and articles prepared for use by newspapers and magazines
- Audio/video presentations on tape, CD, or other electronic media
- Websites, seminars, and other public presentations to audiences
- Direct consultation with individuals
- Elements of worker training programs

The choice of means depends on the specific audience as well as on the specific topic. The choice need not be mutually exclusive.

A brochure, for example, can be printed for distribution by mail or in public places while also being posted on a website, adapted for the text of a press release or periodical article, used as the basis for an informative talk at group meetings, or reformatted as an outline for a video presentation. It might be left (as a brochure) with individuals after one-on-one consultations or given out as part of a worker training program. Finally, it might be simplified and condensed down to key points for a sign or poster.

---

answers), depending on the scope of the remediation and the level of occupant interest. Tell the occupants about the size of the project, planned activities, and remediation timetable. Send or post regular updates on the remediation progress, and send or post a final memo when the project is completed or hold a final meeting.

Try and resolve issues and occupant concerns as they come up. When building-wide communications are frequent and open, those managing the remediation can direct more time toward resolving the problem and less time to responding to occupant concerns.

▶ COMMUNICATE WHEN YOU REMEDIATE

Establish that the health and safety of building occupants are top priorities. Demonstrate that the occupants' concerns are understood and taken seriously. Present clearly the current status of the investigation or remediation efforts. Identify a person whom building occupants can contact directly to discuss questions and comments about the remediation activities. If possible, remediation activities should be scheduled during off-hours when building occupants are less likely to be affected. Communication is important if occupants are relocated during remediation.

The decision to relocate occupants should consider the size of the area affected, the extent and types of health effects exhibited by the occupants, and the potential health risks associated with debris and activities during the remediation project. When considering the issue of relocation, be sure to inquire about, accommodate, and plan for individuals with asthma, allergies, compromised immune systems, and other health-related concerns. Smooth the relocation process and give occupants an opportunity to participate in resolution of the problem by clearly explaining the disruption of the workplace and work schedules. Notify individuals of relocation efforts in advance, if possible.

▶ SAMPLING

Is sampling for mold needed? In most cases, if visible mold growth is present, sampling is unnecessary. In specific instances, such as cases where litigation is involved, the source(s) of the mold contamination is unclear, or health concerns are a problem, you may consider sampling as part of your site evaluation. Surface sampling may also be useful to determine if an area has been adequately cleaned or remediated. Sampling should be done only after developing a sampling plan that includes a confirmable theory regarding suspected mold sources and routes of exposure. Figure out what you think is happening and how to prove or disprove it before you sample! Box 5.6 contains some safety tips.

If you do not have extensive experience and/or are in doubt about sampling, consult an experienced professional. This individual can

---

Box 5.6 Safety Tips While Investigating and Evaluating Mold and Moisture Problems

- Do not touch mold or moldy items with bare hands.
- Do not get mold or mold spores in your eyes.
- Do not breathe in mold or mold spores.
- See later in this chapter for PPE and containment guidelines.
- Consider using PPE when disturbing mold. The minimum PPE is an N95 respirator, gloves, and eye protection.

help you decide if sampling for mold is useful and/or needed, and will be able to carry out any necessary sampling. It is important to remember that the results of sampling may have limited use or application. Sampling may help locate the source of mold contamination, identify some of the mold species present, and differentiate between mold and soot or dirt.

Pre- and post-remediation sampling may also be useful in determining whether remediation efforts have been effective. After remediation, the types and concentrations of mold in indoor air samples should be similar to what is found in the local outdoor air. Since no EPA or other federal threshold limits have been set for mold or mold spores, sampling cannot be used to check a building's compliance with federal mold standards.

Sampling for mold should be conducted by professionals with specific experience in designing mold sampling protocols, sampling methods, and interpretation of results. Sample analysis should follow analytical methods recommended by the American Industrial Hygiene Association (AIHA), the American Conference of Governmental Industrial Hygienists (ACGIH), or other professional guidelines. Types of samples include air samples, surface samples, bulk samples (chunks of carpet, insulation, wall board, etc.), and water samples from condensate drain pans or cooling towers.

A number of pitfalls may be encountered when inexperienced personnel conduct sampling. They may take an inadequate number of samples, there may be inconsistency in sampling protocols, the samples may become contaminated, outdoor control samples may be omitted, and you may incur costs for unneeded or inappropriate samples. Budget constraints will often be a consideration when sampling; professional advice may be necessary to determine whether it is possible to take sufficient samples to characterize a problem on a given budget. If it is not possible to sample properly, with a sufficient number of samples to answer the question(s) posed, it would be preferable not to sample. Inadequate sample plans may generate misleading, confusing, and useless results.

Keep in mind that air sampling for mold provides information only for the moment in time in which the sampling occurred, much like a snapshot. Air sampling will reveal, when properly done, what was in the air at the moment when the sample was taken. For someone without experience, sampling results will

> ## Box 5.7 How to Know When Remediation/Cleanup Is Finished
>
> - You must have completely fixed the water or moisture problem.
> - You should complete mold removal. Use professional judgment to determine if the cleanup is sufficient. Visible mold, mold-damaged materials, and moldy odors should not be present.
> - If you have sampled, the kinds and concentrations of mold and mold spores in the building should be similar to those found outside, once cleanup activities have been completed.
> - You should revisit the site(s) shortly after remediation, and it should show no signs of water damage or mold growth.
> - People should be able to occupy or reoccupy the space without health complaints or physical symptoms.
> - Ultimately, this is a judgment call; there is no easy answer.

be difficult to interpret. Experience in interpretation of results is essential. Box 5.7 lists some things to remember.

▶ PLAN THE REMEDIATION BEFORE STARTING THE WORK

Assess the size of the mold and/or moisture problem and the type of damaged materials before planning the remediation work. See Figure 5.1 for a checklist and Table 5.1 for guidelines. Select a remediation manager for medium or large jobs (or small jobs requiring more than one person). The remediation plan should include steps to fix the water or moisture problem, or the problem may reoccur (see Box 5.8). The plan should cover the use of appropriate personal protective equipment (PPE) and include steps to carefully contain and remove moldy building materials to avoid spreading the mold. A remediation plan may vary greatly depending on the size and complexity of the job, and may require revision if circumstances change or new facts are discovered.

The remediation manager's highest priority must be to protect the health and safety of the building occupants and remediators. It is also important to communicate with building occupants when mold problems are identified. In some cases, especially those involving large areas of contamination, the remediation plan may include temporary relocation of some or all of the building occupants.

The decision to relocate occupants should consider the size and type of the area affected by mold growth, the type and extent of health effects reported by the occupants, the potential health risks that could be associated with debris, and the amount of disruption likely to be caused by remediation activities. If possible,

Checklist for Mold Remediation

Investigate and Evaluate Moisture and Mold Problems
- ☐ Assess size of moldy area (square feet)
- ☐ Consider the possibility of hidden mold
- ☐ Clean up small mold areas and fix moisture problems before they become large problems
- ☐ Select remediation manager for medium- or large-size mold problems
- ☐ Investigate areas associated with occupant complaints
- ☐ Identify sources or cause of water or moisture problems
- ☐ Note type of water-damaged materials (wallboard, carpet, etc.)
- ☐ Check inside air ducts and air-handling units
- ☐ Throughout process, consult qualified professional as necessary

Communicate with Building Occupants at all Stages of Process, as Appropriate
- ☐ Designate contact person for questions and comments about medium- or large-scale remediation as needed

Plan Remediation
- ☐ Adapt or modify remediation guidelines to fit situation; use professional judgment
- ☐ Dry wet, nonmoldy materials within 48 hours to prevent mold growth
- ☐ Select cleanup methods for moldy items
- ☐ Select PPE to protect remediators
- ☐ Select containment equipment to protect building, occupants
- ☐ Select personnel with the experience and training needed to implement remediation plan and use PPE and containment, as appropriate

Remediate Moisture and Mold Problems
- ☐ Fix moisture problem, implement repair and/or maintenance plan
- ☐ Dry wet, nonmoldy materials within 48 hours to prevent mold growth
- ☐ Clean and dry moldy materials
- ☐ Discard moldy porous items that cannot be cleaned

Questions to Consider Before Remediating
- ☐ Are there existing moisture problems in the building?
- ☐ Have building materials been wet more than 48 hours?
- ☐ Are there hidden sources of water?
- ☐ Is the humidity too high (high enough to cause condensation)?
- ☐ Are building occupants reporting musty or moldy odors?
- ☐ Are building occupants reporting health problems?
- ☐ Are building materials or furnishings visibly damaged?
- ☐ Has maintenance been delayed or maintenance plan been altered?
- ☐ Has building been recently remodeled and/or has its use changed?
- ☐ Is consultation with medical or health professionals indicated?

Avoid exposure to and contact with mold
- ☐ Use personal protective equipment (PPE)

Figure 5.1 Checklist for mold remediation.

TABLE 5.1 Guidelines for Remediating Building Materials with Mold Growth Caused by Clean Water

MATERIAL OR FURNISHING AFFECTED	CLEANUP METHODS	PERSONAL PROTECTIVE EQUIPMENT	CONTAINMENT
SMALL—Total surface area affected < 10 ft²			
Books and papers	3	Minimum	None required
Carpet and backing	1, 3	N95 respirator, gloves, and goggles	
Concrete or cinder block	1, 3		
Hard surface, porous flooring (linoleum, ceramic tile, vinyl)	1, 2, 3		
Nonporous, hard surfaces (plastics, metals)	1, 2, 3		
Upholstered furniture and drapes	1, 3		
Wallboard (drywall and gypsum board)	3		
Wood surfaces	1, 2, 3		
MEDIUM—Total surface area affected between 10 and 100 ft²			
Books and papers	3	Limited or Full	Limited
Carpet and backing	1, 3, 4	Use professional judgment, consider potential for remediator exposure and size of contaminated area	Use professional judgment, consider potential for remediator/occupant exposure and size of contaminated area
Concrete or cinder block	1, 3		
Hard surface, porous flooring (linoleum, ceramic tile, vinyl)	1, 2, 3		
Nonporous, hard surfaces (plastics, metals)	1, 2, 3		
Upholstered furniture and drapes	1, 3, 4		
Wallboard (drywall and gypsum board)	3, 4		
Wood surfaces	1, 2, 3		

Continued

TABLE 5.1 Guidelines for Remediating Building Materials with Mold Growth Caused by Clean Water—cont'd

MATERIAL OR FURNISHING AFFECTED	CLEANUP METHODS	PERSONAL PROTECTIVE EQUIPMENT	CONTAINMENT
LARGE—Total surface area affected > 100 ft² or potential for increased occupant or remediator exposure during remediation estimated to be significant			
Books and papers	3	Full	Full
Carpet and backing	1, 3, 4	Use professional judgment, consider potential for remediator/occupant exposure and size of contaminated area	Use professional judgment, consider potential for remediator exposure and size of contaminated area
Concrete or cinder block	1, 3		
Hard surface, porous flooring (linoleum, ceramic tile, vinyl)	1, 2, 3, 4		
Nonporous, hard surfaces (plastics, metals)	1, 2, 3		
Upholstered furniture and drapes	1, 2, 4		
Wallboard (drywall and gypsum board)	3, 4		
Wood surfaces	1, 2, 3, 4		

> ### Box 5.8 Key Steps in Mold Remediation
>
> - Consult health professional as appropriate throughout process
> - Select remediation manager
> - Assess size of mold problem and note type of mold-damaged materials
> - Communicate with building occupants throughout process as appropriate to situation
> - Identify source or cause of water or moisture problem
> - Plan remediation, adapt guidelines to fit situation
> - Select PPE
> - Select containment equipment
> - Select remediation personnel or team
> - Choose between outside expertise or in-house expertise
> - Remediate
> - Fix water or moisture problem
> - Clean and dry moldy materials
> - Discard moldy items that can't be cleaned
> - Dry nonmoldy items within 48 hours
> - Check for return of moisture and mold problem
> - If hidden mold is discovered, reevaluate plan

remediation activities should be scheduled during off-hours when building occupants are less likely to be affected.

Remediators, particularly those with health-related concerns, may wish to check with their doctors or health-care professionals before working on mold remediation or investigating potentially moldy areas. If you have any doubts or questions, you should consult a health professional before beginning a remediation project. Box 5.9 contains some questions to consider before remediation begins.

> ### Box 5.9 Establishing a Remediation Plan
>
> - Are there existing moisture problems in the building?
> - Have building materials been wet more than 48 hours?
> - Are there hidden sources of water or is the humidity too high (high enough to cause condensation)?
> - Are building occupants reporting musty or moldy odors?
> - Are building occupants reporting health problems?
> - Are building materials or furnishings visibly damaged?
> - Has maintenance been delayed or the maintenance plan been altered?
> - Has the building been recently remodeled or has building use changed?
> - Is consultation with medical or health professionals indicated?

► HEATING, VENTILATION, AND AIR-CONDITIONING SYSTEM

Do not run the HVAC system if you know or suspect that it is contaminated with mold. If you suspect that it may be contaminated (it is part of an identified moisture problem, for instance, or there is mold growth near the intake to the system), consult EPA's guide *Should You Have the Air Ducts in Your Home Cleaned?* before taking further action.

► REMEDIATION

In some cases, indoor mold growth may not be obvious. It is possible that mold may be growing on hidden surfaces, such as the back side of dry wall, wallpaper, or paneling; the top of ceiling tiles; the underside of carpets and pads, and so on. Possible locations of hidden mold can include pipe chases and utility tunnels (with leaking or condensing pipes), walls behind furniture (where condensation forms), condensate drain pans inside air handling units, porous thermal or acoustic liners inside ductwork, or roof materials above ceiling tiles (due to roof leaks or insufficient insulation).

Some building materials, such as dry wall with vinyl wallpaper over it or wood paneling, may act as vapor barriers, trapping moisture underneath their surfaces and thereby providing a moist environment where mold can grow. You may suspect hidden mold if a building smells moldy, but you cannot see the source, or if you know there has been water damage and building occupants are reporting health problems. Investigating hidden mold problems may be difficult and will require caution when the investigation involves disturbing potential sites of mold growth—make sure to use personal protective equipment. For example, removal of wallpaper can lead to a massive release of spores from mold growing on the underside of the paper. If you believe that you may have a hidden mold problem, you may want to consider hiring an experienced professional. If you discover hidden mold, you should revise your remediation plan to account for the total area affected by mold growth.

► THE KEY TO MOLD CONTROL IS MOISTURE CONTROL

When addressing mold problems, don't forget to take the source of the moisture problem into account, or the mold problem may simply reappear. Remember to check for high humidity and condensation problems as well as actual water leaks, maintenance issues, and HVAC system problems. Protect the health and safety of the building occupants and remediators. Consult a health professional as needed. Use PPE and containment as appropriate when working with mold.

Fix the water or humidity problem. Complete and carry out repair plan if appropriate. Revise and/or carry out maintenance plan if necessary. Revise the remediation plan as necessary, if more damage is discovered during remediation. Continue to communicate with building occupants, as appropriate to the situation. Be sure to address all concerns.

Completely clean up mold and dry water-damaged areas. Select appropriate cleaning and drying methods for damaged/contaminated materials. Carefully contain and remove moldy building materials. Use appropriate PPE. Arrange for outside professional support if necessary.

▶ HEALTH CONCERNS

If building occupants are reporting serious health concerns, you should consult a health professional. In cases in which a particularly toxic mold species has been identified or is suspected, when extensive hidden mold is expected (such as behind vinyl wallpaper or in the HVAC system), when the chances of the mold becoming airborne are estimated to be high, or sensitive individuals (e.g., those with severe allergies or asthma) are present, a more cautious or conservative approach to remediation is required. Always make sure to protect remediators and building occupants from exposure to mold.

▶ CLEANUP METHODS

Mold can eventually cause structural damage to a school or large building if a mold/moisture problem remains unaddressed for a long time. In the case of a long-term roof leak, for example, molds can weaken floors and walls as the molds feed on wet wood. If you suspect that mold has damaged building integrity, you should consult a structural engineer or other professional with expertise in this area.

Table 5.2 contains guidelines for response to clean water damage within 24 to 48 hours to prevent mold growth. These guidelines are for damage caused by clean water. If you know or suspect that the water source is contaminated with sewage,

TABLE 5.2 Guidelines for Response to Clean Water Damage within 24–48 Hours to Prevent Mold Growth

DAMAGED MATERIAL*	ACTIONS
Books and papers	• For nonvaluable items, discard books and papers. • Photocopy valuable/important items, discard originals. • Freeze (in frost-free freezer or meat locker) or freeze-dry.
Carpet and backing—dry within 24–48 hours	• Remove water with water extraction vacuum. • Reduce ambient humidity levels with dehumidifier. • Accelerate drying process with fans.
Ceiling tiles	• Discard and replace.
Cellulose insulation	• Discard and replace.
Concrete or cinder block surfaces	• Remove water with water extraction vacuum. • Accelerate drying process with dehumidifiers, fans, and/or heaters.
Fiberglass insulation	• Discard and replace.
Hard surface, porous flooring (linoleum, ceramic tile, vinyl)	• Vacuum or damp wipe with water and mild detergent and allow to dry, then scrub if necessary. • Check to make sure underflooring is dry; dry if necessary.
Nonporous, hard surfaces (plastics, metals)	• Vacuum or damp wipe with water and mild detergent and allow to dry; scrub if necessary.
Upholstered furniture	• Remove water with water extraction vacuum. • Accelerate drying process with dehumidifiers, fans, and/or heaters. • May be difficult to completely dry within 48 hours. If piece is valuable, you may wish to consult a restoration/water damage professional who specializes in furniture.
Wallboard (drywall and gypsum board)	• May be dried in place if there is no obvious swelling and seams are intact. If not, remove, discard, and replace. • Ventilate the wall cavity, if possible.
Window drapes	• Follow manufacturers' laundering or cleaning instructions.

TABLE 5.2 cont'd

DAMAGED MATERIAL*	ACTIONS
Wood surfaces	• Remove moisture immediately and use dehumidifiers, gentle heat, and fans for drying. (Use caution when applying heat to hardwood floors.) • Treated or finished wood surfaces may be cleaned with mild detergent and clean water and allowed to dry. • Wet paneling should be pried away from wall for drying.

*If a particular item(s) has high monetary or sentimental value, you may wish to consult a restoration/water damage specialist.

Note: If mold growth has occurred or materials have been wet for more than 48 hours, consult Table 2 of the guidelines. Even if materials are dried within 48 hours, mold growth may have occurred. Items may be tested by professionals if there is doubt. Note that mold growth will not always occur after 48 hours; this is only a guideline. The subfloor under the carpet or other flooring material must also be cleaned and dried. See the appropriate section of Table 2 for recommended actions depending on the composition of the subfloor.

Box 5.10 Cleanup Methods

• **Method 1:** Wet vacuum (in the case of porous materials, some mold spores/fragments will remain in the material but will not grow if the material is completely dried). Steam cleaning may be an alternative for carpets and some upholstered furniture.
• **Method 2:** Damp-wipe surfaces with plain water or with water and detergent solution (except wood—use wood floor cleaner); scrub as needed.
• **Method 3:** High-efficiency particulate air (HEPA) vacuum after the material has been thoroughly dried. Dispose of the contents of the HEPA vacuum in well-sealed plastic bags.
• **Method 4:** Discard or remove water-damaged materials and seal in plastic bags while inside of containment, if present. Dispose of as normal waste. HEPA vacuum area after it is dried.

---

or chemical or biological pollutants, then PPE and containment are required by OSHA. An experienced professional should be consulted if you and/or your remediators do not have expertise remediating in contaminated water situations. Do not use fans before determining that the water is clean or sanitary. Boxes 5.10 and 5.11 provide guidance on cleanup methods.

Box 5.11   Clean Surfaces in Adjacent Areas

After the repairs and cleaning have been completed in the affected areas, it is appropriate to HEPA (high-efficiency particulate air filter) vacuum surfaces in adjacent rooms (if not in the entire building) and to wipe down surfaces where vacuuming is not suitable. This step will reduce the number of particles that might otherwise recirculate in the air as a result of movement of building occupants and normal airflow in the building. Vents and HVAC equipment that may have become contaminated should also be cleaned. (In some cases the HVAC system itself may have been the focus of the remediation.) Upholstered furniture and carpets may harbor significant levels of spores and other particles. They may require professional cleaning or repeated HEPA vacuum cleaning to return them to an acceptable (normal) condition. Normal does not mean antiseptic, as some microbial contamination will be found everywhere.

Mold and Paint

Don't paint or caulk moldy surfaces; clean and dry them before painting. Paint applied over moldy surfaces is likely to peel. A variety of mold cleanup methods are available for remediating damage to building materials and furnishings caused by moisture control problems and mold growth. The specific method or group of methods used will depend on the type of material affected.

Vacuums and Damp Wiping

Wet vacuums are vacuum cleaners designed to collect water. They can be used to remove water from floors, carpets, and hard surfaces where water has accumulated. For further details, see Wet Vacuum section in Chapter 3. For information on high-efficiency particulate air vacuums, see HEPA Vacuum section in Chapter 3.

Mold can generally be removed from nonporous (hard) surfaces by damp wiping or scrubbing with water, or water and detergent. Whether dead or alive, mold is allergenic, and some molds may be toxic. It is important to dry these surfaces quickly and thoroughly to discourage further mold growth. Instructions for cleaning surfaces, as listed on product labels, should always be read and followed. Porous materials that are wet and have mold growing on them may have to be discarded. Since molds will infiltrate porous substances and grow on or fill in empty spaces or crevices, the mold can be difficult or impossible to remove completely.

Mold Remediation and Biocides

The purpose of mold remediation is to remove the mold to prevent human exposure and damage to building materials and furnishings. It is necessary to clean up mold contamination, not just to kill the mold. Dead mold is still allergenic, and some dead molds are potentially toxic. The use of a biocide, such as chlorine bleach, is not recommended as a routine practice during mold remediation, although there may be instances where professional judgment may indicate its use (for example, when immune-compromised individuals are present). In most cases, it is not possible or desirable to sterilize an area; a background level of mold spores will remain in the air (roughly equivalent to or lower than the level in outside air). These spores will not grow if the moisture problem in the building has been resolved.

If you choose to use disinfectants or biocides, always be sure to ventilate the area. Outdoor air may need to be brought in with fans. When using fans, take care not to distribute mold spores throughout an unaffected area. Biocides are toxic to humans, as well as to mold. You should also use appropriate PPE and read and follow label precautions. Never mix chlorine bleach solution with cleaning solutions or detergents that contain ammonia; toxic fumes could be produced.

Some biocides are considered pesticides, and some states require that only registered pesticide applicators apply these products in schools. Make sure anyone applying a biocide is properly licensed, if necessary. Fungicides are commonly applied to outdoor plants, soil, and grains as a dust or spray. Examples include hexachlorobenzene, organomercurials, pentachlorophenol, phthalimides, and dithiocarbamates. Do not use fungicides developed for use outdoors for mold remediation or for any other indoor situation.

Remove Damaged Materials and Seal in Plastic Bags

Building materials and furnishings that are contaminated with mold growth and are not salvageable should be double-bagged using 6-mil polyethylene sheeting. These materials can then usually be discarded as ordinary construction waste. It is important to package mold-contaminated materials in sealed bags before removal from the containment area to minimize the dispersion of mold spores throughout the building. Large items that have heavy mold growth should be covered with polyethylene sheeting and sealed with duct tape before they are removed from the containment area.

▶ PERSONAL PROTECTIVE EQUIPMENT

Always use gloves and eye protection when cleaning up mold. If the remediation job disturbs mold and mold spores become airborne, then the risk of respiratory exposure goes up. Actions that are likely to stir up mold include: breakup of moldy porous materials such as wallboard; invasive procedures used to examine or remediate mold growth in a wall cavity; actively stripping or peeling wallpaper to remove it; and using fans to dry items.

The primary function of PPE is to avoid inhaling mold and mold spores and to avoid mold contact with the skin or eyes. The following sections discuss the different types of PPE that can be used during remediation activities. Please note that all individuals using certain PPE equipment, such as half-face or full-face respirators, must be trained, must have medical clearance, and must be fit-tested by a trained professional. In addition, the use of respirators must follow a complete respiratory protection program as specified by the Occupational Safety and Health Administration.

Gloves and Goggles

Gloves are required to protect the skin from contact with mold allergens (and in some cases mold toxins) and from potentially irritating cleaning solutions. Long gloves that extend to the middle of the forearm are recommended. The glove material should be selected based on the type of materials being handled. If you are using a biocide (such as chlorine bleach) or a strong cleaning solution, you should select gloves made from natural rubber, neoprene, nitrile, polyurethane, or PVC. If you are using a mild detergent or plain water, ordinary household rubber gloves may be used. To protect your eyes, use properly fitted goggles or a full-face respirator with HEPA filter. Goggles must be designed to prevent the entry of dust and small particles. Safety glasses or goggles with open vent holes are not acceptable.

Respiratory Protection

Respirators protect cleanup workers from inhaling airborne mold, mold spores, and dust.

- **Minimum:** When cleaning up a small area affected by mold, you should use an N95 respirator. This device covers the nose and mouth, will filter out 95% of the particulates in the air, and is available in most hardware stores. In situations where a full-face respirator is in use, additional eye protection is not required.

- **Limited:** Limited PPE includes use of a half-face or full-face air purifying respirator (APR) equipped with a HEPA filter cartridge. These respirators contain both inhalation and exhalation valves that filter the air and ensure that it is free of mold particles. Note that half-face APRs do not provide eye protection. In addition, the HEPA filters do not remove vapors or gases. You should always use respirators approved by the National Institute for Occupational Safety and Health.
- **Full:** In situations in which high levels of airborne dust or mold spores are likely or when intense or long-term exposures are expected (e.g., the cleanup of large areas of contamination), a full-face, powered air purifying respirator (PAPR) is recommended. Full-face PAPRs use a blower to force air through a HEPA filter. The HEPA-filtered air is supplied to a mask that covers the entire face or a hood that covers the entire head. The positive pressure within the hood prevents unfiltered air from entering through penetrations or gaps. Individuals must be trained to use their respirators before they begin remediation. The use of these respirators must be in compliance with OSHA regulations.

Boxes 5.12, 5.13, 5.14, and 5.15 provide details on establishing the extent of mold damage.

Box 5.12 Key Engineering Controls and Work Practices for Mold Remediation in Small, Isolated Areas

- The work area should be unoccupied; removing people from adjacent spaces is not necessary but is recommended for infants, persons recovering from surgery, immune-suppressed people, or people with chronic inflammatory lung diseases (e.g., asthma, hypersensitivity pneumonitis, and severe allergies).
- Containment of the work area is not necessary.
- Use dust suppression methods (e.g., misting (not soaking) surfaces prior to remediation).
- Clean and/or remove materials, seal materials being removed in plastic bags.
- The work area and areas used by remediation workers for egress should be cleaned with a damp cloth or mop and a detergent solution.
- Leave area clean, dry, and free of visible debris.

Note: This applies to 10 square feet or less of isolated visible mold growth.

Box 5.13 Key Engineering Controls and Work Practices for Mold Remediation in Mid-Size, Isolated Areas

- The work area should be unoccupied; removing people from adjacent spaces is not necessary but is recommended for infants, persons recovering from surgery, immune-suppressed people, or people with chronic inflammatory lung diseases (e.g., asthma, hypersensitivity pneumonitis, and severe allergies).
- Containment of the work area is not necessary. Cover surfaces in the work area that could become contaminated with secured plastic sheets to contain dust and debris and prevent further contamination.
- Use dust suppression methods (e.g., misting (not soaking) surfaces prior to remediation).
- Clean and/or remove materials, seal materials being removed in plastic bags.
- The work area and areas used by remediation workers for egress should be cleaned with a damp cloth or mop and a detergent solution.
- Leave area clean, dry, and free of visible debris.

Note: This applies to 10 to 30 contiguous square feet.

Box 5.14 Key Engineering Controls and Work Practices for Mold Remediation in Large, Isolated Areas

- Develop a suitable mold remediation plan. If abatement procedures are expected to generate a lot of dust (e.g., abrasive cleaning of contaminated surfaces, demolition of plaster walls) or the visible concentration of mold is heavy (i.e., blanket versus patchy coverage) follow the extensive contamination procedures below.
- Consult with industrial hygienists or other environmental health and safety professionals with experience performing microbial investigations and/or mold remediation before beginning remediation.
- The work area and areas directly adjacent to it should be unoccupied.
- Cover surfaces in the work area and adjacent areas that could become contaminated with secured plastic sheets to contain dust and debris and prevent further contamination.
- Seal ventilation ducts/grills in the work area and areas directly adjacent with plastic sheeting.
- Use dust suppression methods (e.g., misting (not soaking) surfaces prior to remediation).
- Clean and/or remove materials, seal materials being removed in plastic bags.
- Work and surrounding areas should be HEPA vacuumed and cleaned with a damp cloth or mop and a detergent solution.
- All areas should be left dry and free from contamination and debris.

Note: This applies to 30 to 100 contiguous square feet.

> ### Box 5.15 Key Engineering Controls and Work Practices for Mold Remediation Where Extensive Contamination Exists
>
> - Develop a suitable mold remediation plan. For remediation of extensive contamination (greater than 100 contiguous square feet in one area), the plan should address: work area isolation, the use of exhaust fans with HEPA filtration, and the design of airlocks/decontamination room.
> - Consult with industrial hygienists or other environmental health and safety professionals with experience performing microbial investigations and/or mold remediation before beginning remediation.
> - The work area should be unoccupied. If the containment practices listed below will keep mold spores from leaving the contained area, then it may not be necessary to remove people from surrounding areas. However, removal is still recommended for infants, persons recovering from surgery, immune suppressed people, or people with chronic inflammatory lung diseases (e.g., asthma, hypersensitivity pneumonitis, and severe allergies).
> - Before beginning work, cover and seal other surfaces in the work area that could become contaminated with mold spores using plastic sheeting and duct tape; this will help contain dust and debris and prevent further contamination.
> - Contain the affected area. Completely isolate the area to be evaluated and remediated from occupied spaces using plastic sheeting, or other particulate barrier, sealed with duct tape. Use air locks at entry/exit points and provide a sealed decontamination room that is connected to the containment where mold remediation workers must remove PPE before exiting.
> - Shut off the HVAC system and seal ventilation ducts/grills in the work area and adjacent areas to prevent the spread of spores.
> - Keep the work area under negative pressure to minimize the spread of spores to adjacent areas. Use an exhaust fan equipped with HEPA filtration to maintain negative pressure.
> - Use dust suppression methods (e.g., misting (not soaking) surfaces prior to remediation).
> - Clean and/or remove materials, seal materials being removed in plastic bags; wipe down or HEPA vacuum the outside surface of the bags of material being removed.
> - Before removing isolation barriers, HEPA vacuum the contained area and the decontamination room and then clean or mop it with a detergent.
> - Leave area clean, dry, and free of visible debris.
>
> *Note*: This applies to areas that have greater than 100 contiguous square feet.

Disposable Protective Clothing

Disposable clothing is recommended during a medium or large remediation project to prevent the transfer and spread of mold to clothing and to eliminate skin contact with mold.

- **Limited**: Disposable paper overalls can be used.
- **Full**: Mold-impervious disposable head and foot coverings, and a body suit made of a breathable material, such as TYVEK®,

should be used. All gaps, such as those around ankles and wrists, should be sealed (many remediators use duct tape to seal clothing).

▶ CONTAINMENT

The purpose of containment during remediation activities is to limit release of mold into the air and surroundings, in order to minimize the exposure of remediators and building occupants to mold. See Boxes 5.16, 5.17, and 5.18 for information on containment. Mold and moldy debris should not be allowed to spread to areas in the building beyond the contaminated site.

The larger the area of moldy material, the greater the possibility of human exposure and the greater the need for containment. In general, the size of the area helps determine the level of containment. However, a heavy growth of mold in a relatively small area could release more spores than a lighter growth of mold in a relatively large area. Choice of containment should be based on professional judgment. The primary object of containment should be to prevent occupant and remediator exposure to mold.

Limited Containment

Limited containment is generally recommended for areas involving between 10 and 100 square feet (ft^2) of mold contamination. The enclosure around the moldy area should consist of a single layer of 6-mil, fire-retardant polyethylene sheeting. The containment

---

Box 5.16 Is Encapsulation an Option?

Encapsulating moldy areas—permanently sealing them off as a substitute for complete removal or effective cleaning of moldy materials such as drywall or interior cavities—is sometimes suggested as an option where other methods are not considered feasible or affordable. Encapsulation might be done through creating an airtight and inviolable seal between the contaminated space and other parts of the building or might take the form of painting over moldy materials after drying and surface cleaning. Encapsulation is not a cleaning method, is not a best practice, and is not ordinarily recommended, although there may be specific exceptional circumstances. The barrier itself can encourage or be subject to new mold growth by trapping moisture. New mold growth can dislodge wallpaper, make paint bubble, and can emit spores and other particles into occupied space through air pressure and eventual deterioration of seals.

Box 5.17 Containment Tips

- **Limited:** Use polyethylene sheeting ceiling to floor around the affected area with a slit entry and covering flap; maintain area under negative pressure with HEPA filtered fan unit. Block supply and return air vents within containment area.
- **Full:** Use two layers of fire-retardant polyethylene sheeting with one airlock chamber. Maintain area under negative pressure with HEPA filtered fan exhausted outside of building. Block supply and return air vents within containment area.
- Always maintain the containment area under negative pressure.
- Exhaust fans to outdoors and ensure that adequate makeup air is provided.
- If the containment is working, the polyethylene sheeting should billow inwards on all surfaces. If it flutters or billows outward, containment has been lost, and you should find and correct the problem before continuing your remediation activities.

Box 5.18 Isolate Affected Areas

Cleaning and repair will stir up and disperse large numbers of particles, and air movement will carry them at least to adjacent rooms (possibly much farther) without isolation. Mold spores are small and easily dispersed into indoor air even on passive air currents. If the area to be cleaned is significant or if the growth is very heavy even in a relatively small area, the area should be sealed off to prevent spores and other particles from moving to other rooms.

The New York City guidelines recommend using plastic sheeting for a mid-size isolated contamination area (10–30 square feet of visible mold growth) or larger. That, however, is a rule of thumb designed for New York City government buildings and government employees doing the work. More caution may be appropriate for homeowners or commercial building personnel. Multiple isolation chambers (air locks) or decontamination chambers are required to allow entry into and exit from the work area while preventing movement of particles outside the work area. The larger and more extensive the contamination, the more need exists for the additional precautions. Supply air and return ducts should be covered and taped and HVAC (heating, ventilating, air-conditioning system) turned off to prevent microbial particulate contamination from being spread into and through the system.

It may be necessary to establish negative air pressure to ensure that particles do not escape from the contaminated rooms into adjacent rooms or ducts. If so, the need for that precaution may itself suggest the need for professional assistance, as the process requires a negative air machine (or machines) and personnel who are trained in the use of that equipment. The larger the affected area and the heavier the contamination, the more appropriate negative air pressure and more elaborate containment procedures become.

should have a slit entry and covering flap on the outside of the containment area. For small areas, the polyethylene sheeting can be affixed to floors and ceilings with duct tape. For larger areas, a steel or wooden stud frame can be erected and polyethylene sheeting attached to it. All supply and air vents, doors, chases, and risers within the containment area must be sealed with polyethylene sheeting to minimize the migration of contaminants to other parts of the building. Heavy mold growth on ceiling tiles may impact HVAC systems if the space above the ceiling is used as a return air plenum. In this case, containment should be installed from the floor to the ceiling deck, and the filters in the air handling units serving the affected area may have to be replaced once remediation is finished.

The containment area must be maintained under negative pressure relative to surrounding areas. This will ensure that contaminated air does not flow into adjacent areas. This can be done with a HEPA-filtered fan unit exhausted outside of the building. For small, easily contained areas, an exhaust fan ducted to the outdoors can also be used. The surfaces of all objects removed from the containment area should be remediated/cleaned prior to removal.

Full Containment

Full containment is recommended for the cleanup of mold-contaminated surface areas >100 ft^2 or in any situation in which it appears likely that the occupant space would be further contaminated without full containment. Double layers of polyethylene should be used to create a barrier between the moldy area and other parts of the building. A decontamination chamber or airlock should be constructed for entry into and exit from the remediation area. The entryways to the airlock from the outside and from the airlock to the main containment area should consist of a slit entry with covering flaps on the outside surface of each slit entry. The chamber should be large enough to hold a waste container and allow a person to put on and remove PPE. All contaminated PPE, except respirators, should be placed in a sealed bag while in this chamber. Respirators should be worn until remediators are outside the decontamination chamber. PPE must be worn throughout the final stages of HEPA vacuuming and damp-wiping of the contained area. PPE must also be worn during HEPA vacuum filter changes or cleanup of the HEPA vacuum.

► EQUIPMENT

Moisture meters may be helpful for measuring the moisture content in a variety of building materials following water damage. They can also be used to monitor the process of drying damaged materials. These direct reading devices have a thin probe which can be inserted into the material to be tested or pressed directly against the surface of the material. Moisture meters can be used on materials such as carpet, wallboard, wood, brick, and concrete.

Humidity meters can be used to monitor humidity indoors. Inexpensive models are available that monitor both temperature and humidity. A humidistat is a control device that can be connected to the HVAC system and adjusted so that, if the humidity level rises above a set point, the HVAC system will automatically come on.

Use high-quality filters in your HVAC system during remediation. Consult an engineer for the appropriate efficiency for your specific HVAC system and consider upgrading your filters if appropriate. Conventional HVAC filters are typically not effective in filtering particles the size of mold spores. Consider upgrading to a filter with a minimum efficiency of 50 to 60% or a rating of MERV 8, as determined by Test Standard 52.2 of the American Society of Heating, Refrigerating, and Air-Conditioning Engineers. Remember to change filters regularly and change them following any remediation activities.

► STATE REQUIREMENTS

State requirements for abatement can vary. Some of the states may demand procedures that are not included in OSHA guidelines. Mold abatement is not as actively regulated as the abatement of lead or asbestos, but there are still rules and recommendations that should be followed. To expand on this we will now review the general requirements as they are set forth in the state of Connecticut.

The following basic tenets listed should be followed whenever there is a concern about mold.

- Find the source of the water.
- Stop the water and fix any leaks.
- Remove moldy materials.

- Discard porous materials that have remained wet for over 48 hours.
- Perform mold abatement as appropriate, according to an abatement plan designed for the particular site.

The *Connecticut Guidelines for Mold Abatement Contractors* refer to other guidelines for much of the detailed recommendations for mold remediation. These other guidelines have been written for building engineers, property managers, school officials, abatement contractors, and the public. They include guidelines written by the New York City Department of Health and Mental Hygiene, U.S. Environmental Protection Agency, the Canada Mortgage and Housing Corporation, Health Canada, and the Institute of Inspection Cleaning and Restoration Certification (IICRC).

Professional Involvement
There are a number of types of professional contractors that perform various kinds of environmental abatement. These include environmental contractors such as those who perform lead, radon, and/or asbestos abatement, and restoration and cleaning contractors involved in water restoration, smoke and fire restoration, and carpet cleaning specialists. Before entering into the mold abatement business, any contractor should acquire professional training in this area, and become familiar with technical and reference materials referenced in this document.

Training
Contractors should be trained to use state-of-the-art techniques when performing mold abatement to keep building occupants and their own workers safe. The State of Connecticut does not offer training or licensure for mold abatement contractors. There are a number of professional organizations and trade groups that have created credentialing and standards-setting programs in order to "self-police" the industry in the absence of federal and state regulations. The following sections mention some of the national organizations offering training and credentialing in mold abatement. It should be noted that the quality of the training one receives might vary greatly depending on the organization sponsoring the training, the curricula, and the actual trainer.

One organization for contractors, the Institute for Inspection, Cleaning and Restoration Certification (IICRC), has published a voluntary standard for professional mold remediators called *IICRC Standard and Reference Guide for Professional Mold Remediation, S520* (first published in December 2003). This standard is still being revised. The expected release timeframe for the second edition was 2008. The *CT Guidelines for Mold Abatement Contractors* refers readers to the most current version available of *IICRC S520* for a detailed discussion about the technical aspects of mold abatement that space does not allow for here. The CT DPH recommends that as a minimum competency, all mold abatement contractors doing business in Connecticut follow the principles and practices stated in the most current version of *IICRC S520*.

Further, we recommend that every mold abatement jobsite have a full-time supervisor at the jobsite who is formally trained to understand the principles and practices described in *IICRC S520*. We recommend that all workers other than the jobsite supervisor be adequately trained so that they understand the proper use of PPE, know how and when to use such equipment, and can work in a safe manner without causing harm to themselves, fellow workers and building occupants, or the building.

Third-Party Oversight

Third-party oversight should be used whenever the mold abatement contractor has questions about how to abate a certain area, whether certain methods should be employed, or if unusual circumstances exist such as health considerations of building occupants, or questions about how much of an area should be abated (i.e., half of the wall vs. the entire wall). Some specific examples of when third-party oversight is recommended are:

- During mold abatement project(s) in a hospital, nursing home, rehabilitation facility, or medical clinic
- In any setting where there are immune-compromised persons
- Where there has been raw sewage contamination and a determination must be made about what can be salvaged
- Where an independent assessment is called for
- Where it is determined that it would be beneficial to collect samples based on a hypothesis generated from a site assessment.

Note: *The Industrial Hygienist/Indoor Environmental Professional (IH/IEP) should consult with their lab prior to going to the site for instructions on how the lab wants the samples collected and transported.*

Laboratories should be accredited by the American Industrial Hygiene Association and be a current, successful participant in the Environmental Microbiology Laboratory Accreditation Program (EMLAP). EMLAP is specifically designed for labs identifying microorganisms commonly detected in air, fluids, and bulk samples during indoor air quality studies in a variety of settings. Participation assists the laboratory in maintaining high quality standards. If microbiological samples are to be collected and interpreted for the client, this should be performed by the IH/IEP and not the mold abatement contractor.

CT DPH recommends that all individuals acting as consultants on mold abatement projects, whether they are industrial hygienists and/or independent environmental professionals, obtain training regarding indoor air quality and sampling for and interpretation of bacteria and mold in indoor environments. The American Industrial Hygiene Association (AIHA; *www.aiha.org*), the American Conference of Governmental Industrial Hygienists (ACGIH; *www.acgih.org*), and the American Indoor Air Quality Council (*www.iaqcouncil.org*) are some of the national organizations providing this type of training.

Environmental Assessment

The first step in most mold investigations should be to take a building history. Answers to important questions about the age of the building and of the roof, construction history, history of water damage, leaks/floods, and maintenance history helps the IH/IEP and mold abatement contractor gain a larger picture about the building, and may impact sampling and/or remediation strategies.

During the initial walkthrough, emphasis should be placed on looking for evidence of water damage, because this is where mold is likely to be found. Taking note of musty, moldy odors is often a good way to begin an investigation and head toward the direction of the source of the odor (mold). As the basic tenets note, the source of the water must be found and stopped *immediately*, moldy materials must be removed and replaced (if porous), the area must be dried, and abatement must take place according to these guidelines.

Decisions about appropriate remediation strategies are not always reliably made based on visual inspection alone. Mold may be growing in places that are not readily observable, such as on the reverse side of wallboard panels, inside of wall cavities, and inside of HVAC systems. This is what is often referred to as "hidden mold." Hidden mold should be remediated along with mold that is readily visible. Other tools in addition to our senses of sight and smell may be employed to alert the mold abatement contractor to the presence of hidden mold. Some of these are discussed briefly next.

Visual Inspection

Visual inspection of the property may be conducted by the mold abatement contractor, third-party consultant (IH/IEP), or both. According to the New York City "Guidelines on Assessment and Remediation of Fungi in Indoor Environments," visual inspection is the most important initial step in identifying a possible contamination problem. The extent of any water damage and mold growth should be visually assessed. This assessment is important in order to determine remedial strategies. Ventilation systems should also be visually checked, particularly for damp filters but also for damp conditions elsewhere in the system and for overall cleanliness. Ceiling tiles, gypsum wallboard (Sheetrock®), wallpaper, cardboard, paper, and other cellulosic surfaces should be given careful attention during a visual inspection. The use of equipment, such as a boroscope, to view spaces in ductwork or behind walls, or a moisture meter—to detect moisture in building materials—may be helpful in identifying hidden sources of fungal growth and the extent of damage by water. Thermal imaging can also be used to look for gaps in insulation and can be useful in predicting areas in the building where condensation will likely form and areas susceptible to freeze/thaw cycles, and so on. If dampness/high humidity is a possible cause, a hygrometer should be used to measure relative humidity.

Bulk/Surface Sampling

Bulk or surface sampling is not always necessary, and should not be done indiscriminately. It should only be undertaken when the IH/IEP has a hypothesis or theory that sampling results will help him/her answer. Such sampling is seldom needed for relatively small jobs such as those in homes and other residential settings. If samples are collected, it is important that the right kind of sample is collected and analyzed with the appropriate method that will yield meaningful results. Therefore, be sure to follow the following caveats.

- Do not collect samples if you don't know how to interpret the results.
- Do not collect samples if the results will not add any useful information.
- If you do collect samples, always call an AIHA EMLAP accredited lab first to discuss the type of information that you hope to gain, and check to make sure that the particular sampling method you are planning to use will give you what you are looking for.
- Do not sample if the results will not affect your remediation plan.

For more information about testing, refer to the Connecticut DPH fact sheet, "Indoor Air Quality Testing Should Not Be Your First Move."

Air Monitoring

The same caveats for bulk and surface samples apply to collecting air samples for fungi. Samples should not be collected indiscriminately, but only by an IH/IEP to support or refute a hypothesis. Such sampling is seldom needed in homes or other residential settings. Professional judgment based on experience and training should guide the IH/IEP who uses air sampling judiciously as a tool. Communication with an AIHA EMLAP certified laboratory is highly recommended *before* samples are collected. This is to ensure that the proper collection and analytical methods will be used on a particular job to yield meaningful results for the project.

There are no standards for comparison with any air samples collected for microbial agents. Data must be evaluated by an IH/IEP within the context of the entire investigation. Decisions about whether to clean/remediate are almost never made based solely on air monitoring data.

Air monitoring may be useful if presence of mold is suspected (e.g., musty odors) but cannot be identified by visual inspection due to hidden mold, or colorless or light-colored fungi which may not be visible to the naked eye, or bulk sampling (i.e., bulk sampling for hidden mold). The purpose of such air monitoring is to aid in determining the location and/or extent of contamination.

Air monitoring may be necessary if there is evidence from a visual inspection or bulk sampling that ventilation systems may be contaminated. The purpose of such air monitoring is to assess the extent of contamination throughout a building. It is preferable to conduct sampling while ventilation systems are operating.

If air monitoring is performed, for comparative purposes, outdoor air samples should be collected concurrently at a building supply air intake if possible, and at a location representative of outdoor air. For additional information on air sampling, refer to the book by the American Conference of Governmental Industrial Hygienists, *Bioaerosols: Assessment and Control.*

Post-remediation sampling results are almost never zero. This is because mold is everywhere. Even after a thorough cleaning, some mold normally found outdoors will likely migrate into the area as soon as the remediation is complete. The goal of remediation stated in *IICRC S520* is to return a condition 2 or 3 site to that of condition 1 (normal fungal ecology).

Analysis of Environmental Samples

CT DPH recommends the use of laboratories accredited by the American Industrial Hygiene Association's Environmental Microbiology Laboratory Accreditation Program to analyze viable and nonviable air samples and bulk/surface samples for bacteria and fungi in indoor environments. Participation in EMLAP ensures that the lab uses documented quality controlled procedures, and participates in quarterly proficiency testing.

There are some specialty procedures that the IH/IEP may choose, because he/she feels that they may add information that cannot be obtained from culturable or spore trap sampling, or bulk or swab sampling. Some of these specialty procedures include QPCR (quantitative polymerase chain reaction), ergosterol as a measure of fungal biomass (for large-scale buildings and research projects), and endotoxin (for gram negative bacteria, which may also be present during floods).

Laboratories with a great deal of experience should be used to perform these analytical procedures. They are often university, government, or research laboratories. IH/IEPs desiring to include some of these specialty procedures as part of their sampling plan on mold abatement jobs are encouraged to call CT DPH, Environmental and Occupational Health Assessment Program, Indoor Environmental Quality Unit at 860-509-7740 *prior to the start of the job* to discuss this further.

Remediation

The intent of the *Connecticut Guidelines for Mold Abatement Contractors* is not to write a technical procedures and practices document, but rather, to refer readers to documents containing this technical information, such as the *IICRC S520* and NYC *Guidelines*. We point to differences in approaches, and call the readers' attention to key points that mold abatement contractors should be aware of. The goal for remediation should be:

• To eliminate visible mold
• To reduce hidden mold
• To restore the microbial composition to that normally found in ambient outdoor and nonaffected indoor areas

Factors to Consider When Planning Mold Abatement

Indoor mold problems arise from water problems. Knowing whether the water problems are chronic or a one-time occurrence helps inform how the remediation project will be designed and executed. An interdisciplinary approach is often required to perform mold abatement. For example, heating, ventilation, and air-conditioning (HVAC) engineers, hospital infection control practitioners, and facilities engineers may all be a part of the remediation planning team, depending on the setting.

Note that other types of abatement work—for example, lead and asbestos abatement—often include water sprays and/or misting for dust control. However, on mold abatement projects, an important goal is to dry out the environment to prevent mold from propagating. Professional judgment must be used based on training and experience when deciding on the best method(s) for dust control on a mold abatement job, but consideration should be given to HEPA vacuuming in place of, or in conjunction with, the judicious use of light misting to suppress dust in the work area. Refer to the most recent edition of *IICRC S520* for further information.

Determining Scope of Work

The most common method for determining the scope and safety precautions to be used in a mold abatement project is to estimate the size of the contaminated area by visual inspection. This method is used in both the NYC and EPA Guidelines. The "size of contamination" concept has gained popular acceptance because it is easy to understand and communicate, and is workable if the water damage is stopped and handled right away, within the first two days. However, if the water is not stopped and the property is not dried out thoroughly within 48 hours, mold will grow, and can infest porous surfaces. The problem with the "size alone" concept is several-fold:

- It does not take into account the possible presence of hidden mold. If the water caused a significant amount of damage, and the area remained wet/damp for more than 48 hours, there is a good likelihood that mold is growing in both places that are visible and behind/inside visible areas (such as wall cavities, attics, crawl spaces, etc.).
- The "size alone" concept does not take into account special populations/settings such as hospitals, nursing homes, rehabilitation facilities, same-day surgery centers or other medical treatment facilities, or private residences where a chronically debilitated patient lives.

Rather than relying on "size of contamination" for project planning, the *IICRC S520* [voluntary] *Standard and Reference Guide for Professional Mold Remediation* uses the terms *Conditions 1, 2, and 3* to define indoor environments relative to mold. A comparison of this approach with the size of contamination approach described in the NYC *Guidelines* is presented in Box 5.19

You have to balance your approach to mold remediation and abatement. The primary sources of requirements and information needed will come from EPA, OSHA, and your local authorities. Become involved and get to know these regulations before you embark on working with mold.

Box 5.19 IICRC S520 NYC Guidelines

(Note that the NYC "Guidelines on Assessment and Remediation of Fungi in Indoor Environments" refers to active growth.)

Condition 1 (normal fungal ecology): An indoor environment that may have settled spores, fungal fragments, or traces of actual growth with identity, location, and quantity reflective of normal fungal ecology for a similar indoor environment

Condition 2 (settled spores): An indoor environment that is primarily contaminated with settled spores that were dispersed directly or indirectly from a Condition 3 area, and which may have traces of actual growth

Condition 3 (actual growth): An indoor environment that is primarily contaminated with the presence of actual mold growth and associated spores; actual growth includes growth that is active or dormant, visible, or hidden

Level I: Small isolated areas (10 sq. ft or less); e.g., ceiling tiles, small areas on walls

Level II: Mid-sized isolated areas (10–30 sq. ft.); e.g., individual wallboard panels

Level III: Large isolated areas (30–100 square feet); e.g., several wallboard panels

Level IV: Extensive contamination (> 100 contiguous square feet in an area)

SPECIFICS

Although the majority of technical procedures and practices are covered in *IICRC S520* that the Connecticut DPH recommends mold abatement contractors follow, there are several specifics in the guidelines that warrant special attention:

- When performing structural remediation, the contaminated area must be isolated from noncontaminant areas to prevent cross-contamination. This usually involves building a barrier or containment structure, usually made with polyethylene sheeting. The containment structure should be checked to make sure that it does not leak, is strong enough to withstand the number of negative air machines that will be placed inside, and if pressure differentials are lost, containment flaps will close so that contaminated materials remain inside of the structure. Experience and training will help guide the contractor regarding size and construction of containment, but be prepared to expand the containment structure when additional mold is found (i.e., hidden mold), and the scope of the project is expanded.
- If abrasive tools are to be used, the abatement contractor should establish HEPA filtered negative air in the workspace. This limits the potential spread of contamination.
- "Physically removing mold contamination is the primary means of remediation." This means that it is not acceptable to simply spray a product over mold to cover it up. Indiscriminant use of antimicrobial products, coatings, sealants, and cleaning chemicals is also not recommended. They may be used as complementary tools on certain surfaces after the mold has been removed.
- Mold resistant coatings/sealants should not be sprayed on top of actively growing mold.

Box 5.19 cont'd

- Fungicidal coatings (those rated to kill mold) should not be used as sealants or encapsulants on active, viable mold.
- The use of antimicrobial agents in the form of fogging agents is not recommended for mold remediation in buildings. These are gas or vaporphase antimicrobials that, by the nature of the delivery system, do not offer enough concentration and contact time to be effective at killing mold. Other problems include toxicity, inefficient capture rate, and the fact that physical removal is still necessary after fogging.
- Biocides are useful in treating indoor environments flooded with raw sewage.

REMEDIATION OF HVAC SYSTEMS

IICRC S520, Section 11 is devoted to the remediation of HVAC systems. This section refers to the National Air Duct Cleaners Association's (NADCA) document, *ACR 2006: Assessment, Cleaning, and Restoration of HVAC Systems*. This document has become an industry standard. CT DPH recommends that mold abatement contractors engaged in remediation of HVAC systems follow S520 Section 11 (or appropriately renumbered section in future revisions) document thoroughly, and refer to NADCA ACR 2006 for further technical information.

As mentioned before, the majority of technical procedures and practices are covered in the *IICRC S520* that CT DPH recommends mold abatement contractors follow. However, there are a few key points in S520 that warrant special attention:

- Isolating HVAC components from other parts of the building that are undergoing abatement is recommended.
- Use of biocides in HVAC systems is discouraged. If there is a fungal infestation inside of ducts or other HVAC components, the mold must be physically removed. It is not appropriate to spray antimicrobial products, coatings, sealants, or encapsulants on top of viable or nonviable mold in HVAC systems.
- Use of coatings and sealants prophylactically (during new installation or new construction) or as a final treatment post-remediation as a lockdown for residual particulate and to provide a smooth, clean surface to deter future fungal activity may be of some value.

HEALTH AND SAFETY, HAZARD COMMUNICATION

IICRC S520 refers to pertinent OSHA standards for occupational health concerns, including the OSHA General Duty Clause, Emergency Action and Fire Protection Plans, PPE, Respiratory Protection, Asbestos, Lead, Heat Disorders and Health Effects, Confined Spaces, Hazard Communication, Lockout/Tagout, Fall Protection, Noise Protection, and Scaffolds. All of the OSHA standards should be followed during any remediation work.

Contractors who get involved in mold abatement are likely to encounter asbestos and lead on the job at some point. CT DPH has specific regulations for each of these substances. Mold abatement contractors must follow all state regulations pertaining to asbestos and lead. To obtain further information and contacts within CT DPH, the main web address is *www.ct.gov/dph*.

Mold Following Hurricanes and Floods

6

Mold that develops after hurricanes and floods can cause health concerns and significant financial losses. When water enters a living space it has to be removed and dried out quickly to prevent mold from prospering in the environment. The spores of mold can wreak havoc on the health of some people. In fact, the health effects associated with mold can force some people from their homes.

The cleanup process after water infiltration can become quite expensive. Mold growing in a damp bathroom is not usually difficult to control, but when a hurricane or flood inundates a building with water the remediation process changes and the cost skyrockets. Most people don't give much thought to mold until they suffer from the potential health effects or see the nasty growth infesting their home, office, or school. Then they think about it. If you are one of the contractors who will be called to deal with major mold problems, you need to be prepared. That is the purpose of this chapter, so let's get to it.

▶ FUNGI

Molds, mushrooms, mildews, and yeasts are all classified as *Fungi*, a kingdom of organisms distinct from plants and animals. Fungi differ from plants and animals in several respects. Unlike animals, fungi have cell walls. However, unlike plants, which also have cell walls, fungal cell walls are made mostly of chitin and glucan. Fungi cannot produce their own nutrients as plants do through photosynthesis. Fungi secrete enzymes that digest the material in which the fungi are embedded and absorb the released nutrients. Multicellular fungi do not differentiate into different organs or functional components the way plants and animals do.

It is known that approximately 100,000 species of fungi exist; fewer than 500 fungal species have been described as human pathogens that can cause infections. Visible growth of multicellular fungi consisting of branching filamentous structures (mycelia) are popularly known as molds, and are referred to by that term in this chapter.

Molds are ubiquitous in nature and grow almost anywhere indoors or outdoors. The overall diversity of fungi is considerable. For example, the genus *Aspergillus* has at least 185 known species. Molds spread and reproduce by making spores, which are small and lightweight, able to travel through air, capable of resisting dry, adverse environmental conditions, and capable of surviving a long time. The filamentous parts of mold (hyphae) form a network called mycelium, which is observed when a mold is growing on a nutrient source. Although these mycelia are usually firmly attached to whatever the mold is growing on, they can break off, and persons can be exposed to fungal fragments. Some micro-organisms, including molds, also produce characteristic volatile organic compounds (VOCs) or microbial VOCs (MVOCs). Molds also contain substances known as beta glucans; MVOCs and beta glucans might be useful as markers of exposure to molds.

Some molds are capable of producing toxins (sometimes called mycotoxins) under specific environmental conditions, such as competition from other organisms or changes in the moisture or available nutrient supply. Molds capable of producing toxins are popularly known as toxigenic molds; however, use of this term is discouraged because even molds known to produce toxins can grow without producing them. Many fungi are capable of toxin production, and different fungi can produce the same toxin.

▶ FACTORS THAT PRODUCE MOLD GROWTH

Although molds can be found almost anywhere, all of them need moisture and nutrients to grow. The exact specifications for optimal mold growth vary by the species of mold. However, mold grows best in damp, warm environments. The availability of nutrients in indoor environments rarely limits mold growth because wood, wallboard, wallpaper, upholstery, and dust can be nutrient sources. Similarly, the temperature of indoor environments, above freezing and below the temperature for denaturing proteins, can support mold growth, even if the actual temperature is not optimal.

The primary factor that limits the growth of mold indoors is lack of moisture. Substantial indoor mold growth is virtually synonymous with the presence of moisture inside the building envelope. This intrusion of moisture might be from rainwater leaking through faulty gutters or a roof in disrepair, from a foundation leak, from condensation at an interface (e.g., windows or pipes) or between a cold and a warm environment. Water also can come from leaks in the plumbing or sewage system inside the structure. Studies of mold growth on building materials, such as plywood, have found that mold grows on materials that remain wet for 48 to 72 hours. Flooding, particularly when floodwaters remain for days or weeks, provides an almost optimal opportunity for mold growth.

▶ HOW PERSONS ARE EXPOSED TO MOLD

Mold exposure can produce disease in several ways. Inhalation is usually presumed to be the most important mechanism of exposure to viable (live) or nonviable (dead) fungi, fungal fragments or components, and other dampness-related microbial agents in indoor environments. The majority of fungal spores have aerodynamic diameters of 2 to 10 μm, which are in the size range that allow particles to be deposited in the upper and lower respiratory tract. Inhalation exposure to a fungal spore requires that the spore be initially aerosolized at the site of growth. Aerosolization can happen in many ways, ranging from disturbance of contaminated materials by human activity to dispersal of fungi from contaminated surfaces in heating, ventilating, and air-conditioning (HVAC) systems. Fungal spores also can be transported indoors from outdoors. Overall, the process of fungal-spore aerosolization and related issues (e.g., transport, deposition, resuspension, and tracking of fungi to other areas) are poorly understood.

Persons can be exposed to mold through skin contact, inhalation, or ingestion. Because of the ubiquity of mold in the environment, some level of exposure is inevitable. Persons can be exposed to mold through contact with airborne spores or through contact with mycelial fragments. Exposure to high airborne concentrations of mold spores could occur when one comes into contact with a large mass of mold, such as might occur in a building that has been flooded for a long time. Exposure to mycelia fragments could occur when a person encounters a nutrient source for mold that has become disrupted, such as would occur during removal of mold-contaminated building material. Skin contact or exposure by inhalation to either spores or mycelial fragments also could occur in a dusty environment, if the components of dust include these fungal elements.

For the majority of adverse health outcomes related to mold exposure, a higher level of exposure to living molds or a higher concentration of allergens on spores and mycelia results in a greater likelihood of illness. However, no standardized method exists to measure the magnitude of exposure to molds. In addition, data are limited about the relation between the level of exposure to mold and how that causes adverse health effects and how this relation is affected by the interaction between molds and other microorganisms and chemicals in the environment. For this reason, it is not possible to sample an environment, measure the mold level in that sample, and make a determination as to whether the level is low enough to be safe or high enough to be associated with adverse health effects.

Persons affected by major hurricanes or floods probably will have exposure to a wide variety of hazardous substances distributed by or contained within the floodwater. This chapter does not provide a comprehensive discussion of all such potential hazards. Such situations require case by case evaluation and assessment. Guidance has been provided by CDC for such issues in a number of documents, including NIOSH's "Hazard Based Interim Guidelines: Protective Equipment for Workers in Hurricane Flood Response" and CDC's guidance in "Protect Yourself from Chemicals Released during a Natural Disaster."

▶ FACTORS THAT CAUSE DISEASE FROM MOLD

Numerous species of mold cause infection through respiratory exposure. In general, persons who are immune suppressed are at increased risk for infection from mold. Immunosuppression can result from immunosuppressive medication, from medical conditions and diseases that cause immune suppression, or from therapy for cancer that causes transient immune suppression. Although certain species of mold cause infection, many mold species do not cause infection. Infections from mold might be localized to a specific organ or disseminated throughout the body.

Many of the major noninfectious health effects of mold exposure have an immunologic (i.e., allergic) basis. Exposure to mold can sensitize persons, who then might experience symptoms when reexposed to the same mold species. For sensitized persons, hay fever symptoms and asthma exacerbations are prominent manifestations of mold allergy. Although different mold species might have different propensities to cause allergy, available data do not permit a relative ranking of species by risk for creating or exacerbating allergy. In addition, exposure to beta glucans might have an inflammatory effect in the respiratory system.

Prolonged exposure to high levels of mold (and some of the bacterial species) can produce an immune-mediated disease known as hypersensitivity pneumonitis. Clinically, hypersensitivity pneumonitis is known by the variety of exposures that can cause this disorder (e.g., farmer's lung, woodworker's lung, and malt workers' lung).

Ingesting toxins that molds produce can cause disease. Long-term ingestion of aflatoxins (produced by *Aspergillus* species) has been associated with hepatocellular cancer. In addition, ingestion of high doses of aflatoxin in contaminated food causes aflatoxicosis and can result in hepatic failure. Whether concentrations of airborne mold toxins are high enough to cause human disease through inhalation is unknown, and no health effects from airborne exposure to mold-related toxins are proven.

▶ ASSESSING EXPOSURE TO MOLD

Any structure flooded after hurricanes or major floods should be presumed to contain materials contaminated with mold if those materials were not thoroughly dried within 48 hours. In such cases, immediate steps to reduce the risk for exposure to mold are likely to be of greater importance than further exposure assessment steps presented next.

Assessing the level of human exposure to mold in flooded buildings where mold contamination is not obvious is often a central and ongoing activity in recovery related to hurricanes and floods. Understanding the strengths and limitations of the approaches that are available to assess such exposures is important. Buildings that were not flooded could also have mold. For example, buildings with leaking roofs or pipes, which allows water to penetrate into biodegradable building materials, or excessive humidity, particularly in buildings built with biodegradable materials, are susceptible to mold growth.

Visual Inspection and Moisture Assessment

A visual inspection is the most important step in identifying possible mold contamination. The extent of any water damage and mold growth should be visually assessed. This assessment is particularly important in determining remedial strategies and the need for personal protective equipment (PPE) for persons in the contaminated area. Ceiling tiles, gypsum wallboard (Sheetrock), cardboard, paper, and other cellulosic surfaces should be given careful attention during a visual inspection. Not all mold contamination is visible; with a flood, contamination in the interior wall cavities or ceiling is common. A common means of assessing the mold contamination of a building is to estimate the total square feet of contaminated building materials. However, professional judgment will necessarily play an important role in the visual inspection because less quantifiable factors (e.g., location of the mold, building use, and function) and exposure pathways are also important in assessing potential human exposure and health risks.

Ventilation systems also should be visually checked, particularly for damp filters, damp conditions elsewhere in the system, and overall cleanliness. To avoid spreading microorganisms throughout the building, HVAC systems known or suspected to be contaminated with mold should not be run. Guidelines from the U.S. Environmental Protection Agency (EPA) and CDC provide useful information concerning this topic. Different algorithms for assessing and remediating mold-contaminated buildings are available. Examples of such algorithms are available from the U.S. Army, the New York City Department of Health (NYC DPH), and OSHA.

Moisture meters provide qualitative moisture levels in building materials and might be helpful for measuring the moisture content in a variety of building materials (e.g., carpet, wallboard, wood, brick, and concrete) following water damage. Meters also can be used to monitor progress in drying wet materials. Damaged materials should be removed and discarded. Moisture meters are available from contractor tool and supply outlets. Humidity meters can be used to monitor the indoor humidity.

Inexpensive models that monitor both temperature and humidity are available.

A borescope is a hand-held tool that allows users to see hidden mold problems inside walls, ceiling plenums, crawl spaces, and other tight areas. No major drilling or cutting of dry wall is required.

Sampling for Mold

Sampling for mold is not part of a routine building assessment. In most cases, appropriate decisions about remediation and the need for PPE can be made solely on the basis of visual inspection. If visible mold is present, then it should be remediated regardless of what types of microorganisms are present, what species of mold is present, and whether samples are taken. Other than in a controlled, limited, research setting, sampling for biologic agents in the environment cannot be meaningfully interpreted and would not substantially affect relevant decisions about remediation, reoccupancy, handling or disposal of waste and debris, worker protection or safety, or public health. If sampling is being considered, a clear purpose should exist. For example:

- To help evaluate a source of mold contamination. For example, testing the types of mold and mold concentrations indoors versus outdoors can be used to identify an indoor source of mold contamination that might not be obvious on just visual inspection.
- To help guide mold remediation. For example, if mold is being removed and it is unclear how far the colonization extends, then surface or bulk sampling in combination with moisture readings might be useful.

Types of Samples

Types of samples used to assess the presence of mold and the potential for human exposure to mold in a water-damaged building include air samples, surface samples, bulk samples, and water samples from condensate drain pans or cooling towers. Detailed descriptions of sampling and analysis techniques have been published.

Among the types of samples that can be taken, airborne sampling might be a good indicator of exposure from a theoretical point of view, particularly for assessing acute short-term exposures.

However, in practice, many problems (e.g., detection problems and high variability over time) limit the usefulness of these types of samples for most biologic agents. If air sampling is conducted, personal measurements best represent the current exposure, although practical constraints might make personal sampling difficult. Therefore, area sampling is the most commonly performed type of air sampling used to assess bioaerosol exposure despite resultant uncertainty about how accurately the measurements reflect actual personal exposure.

One type of surface sampling is the sampling of settled dust. A theoretical advantage of settled-dust sampling is the presumed correlation of concentrations of fungi in the settled dust with chronic exposure to those fungi. However, surface sampling is a crude measure and will yield a poor surrogate for airborne concentrations. Results of surface sampling as a measure of exposure should be interpreted with caution. Bulk samples can provide information about possible sources of biologic agents in buildings and the general composition and relative concentrations of those biologic agents.

Assessment of Microorganisms

Two distinct approaches are used for evaluation of the presence of specific microbes: culture-based and nonculture-based. The strengths and limitations of the different approaches have been published. Instead of measuring culturable or nonculturable fungi or fungal components, constituents or metabolites of microorganisms can be measured as a surrogate of microbial exposure. Examples of such techniques include polymerase chain reaction (PCR) technologies and immunoassays. Methods for measuring microbial constituents (with some exceptions) are in an experimental phase and have not yet been routinely applied in clinical assessments, risk assessments, or epidemiologic studies.

No health-based standards (e.g., OSHA or EPA standards) or exposure limits (e.g., NIOSH-recommended exposure limits) for indoor biologic agents (airborne concentrations of mold or mold spores) exist. Differences in season; climatic and meteorological conditions; type, construction, age, and use of the building and ventilation systems; and differences in measurement protocols used in various studies (e.g., viable versus nonviable microorganism sampling, sampler type, and analysis) make it difficult to interpret sampling data relative to information from the medical literature. If sampling is performed, exposure data can be evaluated (either quantitatively or qualitatively) by comparing exposure data with background data, indoor environments with outdoor environments, or problem areas with nonproblem areas.

A quantitative evaluation involves comparing exposures, whereas a qualitative evaluation could involve comparing species or genera of microorganisms in different environments. Specifically, in buildings without mold problems, the qualitative diversity of airborne fungi indoors and outdoors should be similar. Conversely, the dominating presence of one or two kinds of fungi indoors and the absence of the same kind outdoors might indicate a moisture problem and degraded air quality.

In addition, the consistent presence of fungi, such as *Stachybotrys chartarum*, *Aspergillus versicolor*, or various *Penicillium* species, over and beyond background concentrations might indicate a moisture problem that should be addressed. Indoor and outdoor mold types should be similar, and indoor levels should be no greater than levels outdoors or in noncomplaint areas. Analytical results from bulk material or dust samples also might be compared with results of similar samples collected from reasonable comparison areas.

Biomarkers

For biologic agents, few biomarkers of exposure or dose have been identified, and their validity for exposure assessment in the indoor environment is often unknown. Testing to determine the presence of immunoglobulin E (IgE) to specific fungi might be a useful component of a complete clinical evaluation in the diagnosis of illnesses (e.g., rhinitis and asthma) that can be caused by immediate hypersensitivity. Testing is usually done by in vitro tests for serum specific IgE, or by skin prick or puncture tests. Detection of immunoglobulin G (IgG) to specific fungi has been used as a marker of exposure to agents that might cause illnesses such as hypersensitivity pneumonitis. However, the ubiquitous nature of many fungi and the lack of specificity of fungal antigens limit the usefulness of these types of tests in evaluating possible building-related illness and fungal exposure.

Specific serologic tests (e.g., tests for cryptococcal antigen, coccidioidal antibody, and *Histoplasma* antigen) are useful in the diagnosis of some fungal infections, but these are the exception. Routine clinical use of immunoassays as a primary means of assessing environmental fungal exposure or health effects related to fungal exposure is not recommended. Health-care providers whose patients express concern about the relation between symptoms and possible exposure to fungi are advised to use immunoassay results with care and only in combination with other clinical information, including history, physical findings, and other laboratory results.

Mycotoxins

In recent years, increased concern has arisen about exposure to specific molds that produce substances called mycotoxins. Health effects related to mycotoxins are generally related to ingestion of large quantities of fungal-contaminated material. No conclusive evidence exists of a link between indoor exposure to airborne mycotoxin and human illness. Many molds can potentially produce toxins given the right conditions. Some molds that produce mycotoxins are commonly found in moisture-damaged buildings; research related to the importance of these findings is ongoing. Although the potential for health problems is an important reason to prevent or minimize indoor mold growth and to remediate any indoor mold contamination, evidence is inadequate to support recommendations for greater urgency of remediation in cases where mycotoxin-producing fungi have been isolated.

The interpretation of environmental sampling data generally requires professional judgment, and medical conclusions cannot be made based solely on the results of analysis of environmental sampling. In the context of mold growth following a major hurricane or flood, mold growth itself and the extent of growth based on a thorough visual inspection is sufficient to categorize a building as moldy or not moldy. This should provide sufficient information for action and no additional characterization is needed.

▶ CLEANUP AND PREVENTION

The most effective way to eliminate mold growth is to remove it from materials that can be cleaned and to discard materials that cannot be cleaned or are physically damaged beyond use. Persons with respiratory conditions, allergies, asthma, or weakened immune systems should avoid mold cleanup if possible or seek the advice of their doctor and determine what type of personal protective equipment is appropriate.

Appropriate PPE (e.g., tight-fitting NIOSH-approved N95 respirator, gloves to limit contact of mold and cleaning solutions with skin, and goggles) should be used when performing cleanup or other activities in mold-contaminated homes or buildings after a flood. See Box 6.1 for information about dealing with contaminated water.

Removing mold problems requires a series of actions. The order of these actions is sometimes important, but might vary on a case-by-case basis. Typically, the following actions are taken regardless of whether a problem is small and simple or large and complex:

- Take emergency action to stop water intrusion, if needed.
- Determine extent of water damage and mold contamination.
- Plan and implement remediation activities.
  - If needed, establish containment and protection for workers and occupants.
  - Eliminate or limit water or moisture sources.
  - Decontaminate or remove damaged materials, as appropriate.
  - Dry any wet materials, if possible.
  - Evaluate whether space has been successfully remediated.
  - Reassemble the space to prevent or limit possibility of recurrence by controlling sources of moisture.

Box 6.1 Key Engineering Controls and Work Practices for Contaminated Water and/or Floodwaters

- Reduce the exposure to splash or aerosolized liquid hazards by limiting the number of people in the area and having those in the area stay upwind of water discharge areas.
- Ensure that good hygiene, especially hand washing, is practiced before eating, drinking, and smoking. If clean water is not available, use an alternative such as hand sanitizer or sanitizing wipes.
- Ensure that cuts and bruises are protected from contact with contaminated water.
- Clean areas of the body that come in contact with contaminated water with soap and water, hand sanitizer, or sanitizing wipes.

ADDITIONAL PERSONAL PROTECTIVE EQUIPMENT

- Goggles if routinely working near splashing floodwater
- N95, R95, or P95 respirators may be necessary for exposure to contaminated water that may become aerosolized
- Watertight boots with steel toe and insoles
- Waterproof gloves for contact with contaminated water

For small, simple problems, the entire list of tasks can be done by one person. Large, complex problems might require many persons from different professions and trades. For circumstances that fall between those extremes, some combination of occupant action and professional intervention will be appropriate. In general, no single discipline brings together all the required knowledge for successful assessment and remediation.

Returning to Mold-Contaminated Homes or Buildings after a Flood

When persons return to homes or buildings after a flood, they should take the following steps:

- Clean up and dry out building quickly. Open doors and windows and use fans or dehumidifiers to dry the building.
- Remove all porous items that have been wet for more than 48 hours and that cannot be thoroughly cleaned and dried. These items can remain a source of mold growth and should be removed from the home or building. Porous, noncleanable items include carpeting and carpet padding, upholstery, wallpaper, drywall, ceiling tiles, insulation material, some clothing, leather, paper, some wood and wood products, and food. Removal and cleaning are important because even dead mold can cause allergic reactions.
- Clean wet items and surfaces with detergent and water to prevent mold growth.
- Temporarily store damaged or discarded items outside the home or building until insurance claims can be processed.

Removing and Cleaning Up Mold in a Building

For cleaning mold covering less than 10 ft^2 in an area flooded by clean water, detergent and water might be adequate. However after hurricanes and major floods, flood water is likely to be contaminated and, in this setting, mold can be removed with a bleach solution of 1 cup chlorine bleach per 1 gallon of water. Never mix bleach or bleach-containing products with ammonia or ammonia-containing products. If water damage is substantial or mold growth covers more than 10 ft^2, consult the EPA guide *Mold Remediation in Schools and Commercial Buildings*.

Some companies specialize in water damage restoration and can assess the issues involved in cleaning up homes after a flood. Two professional trade groups that might be able to help locate such an expert are the Restoration Industry Association (*www.restorationindustry.org*) and the Institute of Inspection, Cleaning, and Restoration Certification (*www.iicrc.org*).

Contractors used for remediation should have experience in cleaning mold. Check references and ask the contractor to follow the recommendations in the guidelines of the American Conference of Governmental Industrial Hygienists (ACGIH) or other guidelines from professional organizations or state agencies. Contact your state health department's website for information about state licensing requirements for contractors in your state. Examples of states that have dealt with natural disasters include Texas and Louisiana.

Cleaning Clothes, Textiles, or Stuffed Animals

Ensure that laundry is washed in safe water. Use only water that is properly disinfected or that the authorities have stated is safe. Take the appropriate steps to make sure that use of gas or electric appliances is safe. Before using a washing machine that was in a flooded building, run the machine through one full cycle before washing clothes. Use hot water and a disinfectant or sanitizer. Take clothes and linens outdoors and shake off any dried mud or dirt before washing them. Hose off muddy items to remove all dirt before putting them in the washer.

If the items are only wet, they can be laundered normally. Check the labels on clothes and linens and wash them in detergent and warm water if possible, or take them to a professional cleaner. Adding chlorine bleach to the wash cycle will remove most mildew and will sanitize the clothing. However, bleach might fade some fabrics and damage other fabrics. If the label reads "dry clean only," shake out loose dirt and take the item to a professional cleaner.

Consult a remediation professional for advice on whether heavily mold-contaminated items made of leather, suede, or a similar material are salvageable or should be discarded. Do not burn or bury textiles that cannot be cleaned. Put them into properly sealed plastic bags and dispose of them as you would normal household garbage in your area.

Salvaging Household Items

When assessing or remediating mold contamination to a house, homeowners or cleanup personnel might decide to repair or clean household items (e.g., housewares or kitchen items) damaged or contaminated by flood waters. As with clothing and other textiles, make sure the water being used is safe. Use only water that is properly disinfected or that the authorities have stated is safe.

Nonporous items (e.g., dishes, pots, glass items, and hard plastic items) can be salvaged. However, because floodwaters are contaminated, nonporous items should be washed by hand in a disinfectant and then air-dried. Do not use a dish towel. Porous items (e.g., cloth, some wood and wood products, and soft plastic) must be discarded because they probably absorbed whatever contaminants were in the floodwaters.

Before using the dishwasher, clean and disinfect it. Then use a hot setting to wash your pots, pans, dishes, and utensils. Do not use the energy-saving setting. Throw away canned foods that are bulging, opened, or damaged. Food containers with screw-caps, snap-lids, crimped caps (soda pop bottles), twist caps, flip tops, snap-open, and home-canned foods should be discarded if they have come into contact with floodwater because they cannot be disinfected. If intact cans have come in contact with floodwater or storm water, remove the labels, wash the cans, and dip them in a solution of 1 cup of bleach in 5 gallons of water. Relabel the cans with a marker.

Cleaning a Heating, Ventilating, and Air-Conditioning System

All surfaces of an HVAC system and all its components that were submerged during a flood are potential reservoirs for dirt, debris, and microorganisms, including bacteria and mold. In addition, moisture can collect in areas of HVAC system components that were not submerged (e.g., air supply ducts above the water line), and this also can lead to the growth of microorganisms. Therefore, all floodwater-contaminated and moisture-laden components of the HVAC system should be thoroughly inspected, cleaned of dirt and debris, and disinfected by a qualified professional. CDC has prepared recommendations for professionals to help ensure that floodwater-contaminated HVAC system components are properly cleaned and remediated. If HVAC systems are not properly cleaned and disinfected to prevent the dissemination of mold and other debris throughout a building, bioaerosols of mold and other microorganisms might exist and can cause a variety of adverse health effects to the building's occupants. Ensure that the HVAC system is shut down before any remedial activities.

Prevention of Mold after the Flood

Limited scientific information exists on the efficacy and impact of prevention strategies. In addition, little of the practical knowledge that has been acquired and applied by design, construction, and maintenance professionals has been subject to thorough validation. No generally accepted health-based standards exist for remediation.

If property owners decide to make extensive repairs or completely rebuild after a flood, they might consider designing and building in a way that will limit the potential for future mold growth. The key to prevention of mold is to eliminate or limit the conditions that foster microbial growth by limiting water intrusion and the nutrients that allow mold to grow. The two basic approaches are to keep moisture-sensitive materials dry and to use materials that are not easily biodegradable or which offer a poor substrate for mold growth.

▶ PERSONAL PROTECTIVE EQUIPMENT

Workers and their employers might be required to wear or to provide protection to minimize exposure to mold. Workers and employers should refer to pertinent OSHA standards and NIOSH guidelines. Information is also provided for the public.

Minimizing exposure to mold involves using PPE and administrative and engineering controls. Administrative controls include identifying mold-contaminated areas promptly, restricting access to these areas, and minimizing aerosol generating activities (e.g., by suppressing dust). Engineering controls include ventilating mold-contaminated areas adequately and using heavy equipment with sealed positive pressure, air-conditioned cabs that contain filtered air recirculation units to protect the workers. Misting contaminated materials with water is a control measure used to reduce dust levels during debris removal.

Workers should wear personal protective equipment regardless of the engineering controls used, especially for skin and eye protection. Primary functions of PPE in a mold-contaminated environment are prevention of the inhalation and ingestion of mold and mold spores and prevention of mold contact with skin or eyes. PPE requirements for workers are likely to differ from the PPE recommendations for homeowners or other building occupants who are less likely to disturb and aerosolize contaminated materials. In addition, PPE recommendations for persons with underlying illness or compromised immune systems will differ from PPE recommendations for healthy persons. Proper training or instruction in the use of protective equipment is essential for effective use. Guidelines for protection of and training recommendations for workers have been published.

Skin and Eye Protection

Gloves keep the hands clean and free from contact with mold. Gloves also protect hands from potentially irritating cleaning solutions. Long gloves that extend to the middle of the forearm are recommended. The glove material should be selected on the basis of the type of substance or chemical being handled. When using a biocide (e.g., chlorine bleach) or a strong cleaning solution, gloves made from natural rubber, neoprene, nitrile, polyurethane, or PVC are needed. When using a mild detergent or plain water, ordinary household rubber gloves can be used. Latex or nonlatex medical examination gloves should be used if hands are likely to be in contact with infectious materials. Persons with natural rubber latex allergy should not use natural rubber latex gloves and should consult the NIOSH Alert on latex gloves for further information.

Properly fitted goggles or a full face-piece respirator are needed to protect eyes. Goggles must be designed to prevent the entry of dust and small particles. Safety glasses or goggles with open vent holes are not appropriate in mold remediation. CDC has published guidelines on this topic.

Protective Clothing

When conducting building inspections and remediation work, workers or homeowners might encounter hazardous biologic agents in addition to chemical and physical hazards. Consequently, appropriate personal protective clothing, either reusable or disposable, is recommended to minimize cross-contamination between work areas and clean areas, to prevent the transfer and spread of mold and other contaminants to street clothing, and to eliminate skin contact with mold or chemicals. In hot environments, precautions to prevent dehydration and heat stress when wearing protective clothing (e.g., drink plenty of water) are needed.

Disposable PPE should be discarded after it is used. Such equipment should be placed into impermeable bags and usually can be discarded as ordinary construction waste. Protective equipment for biocide applicators (e.g., goggles or face shield, aprons or other protective clothing, gloves, and respiratory protection) must be selected on the basis of the product manufacturer's warnings and recommendations. In addition, the manufacturer's recommended precautions should be followed. Reusable protective clothing, including respiratory equipment, should be cleaned according to manufacturers' recommendations for PPE exposed to mold and other potentially hazardous chemicals (e.g., bleach and biocides).

Respiratory Protection

Inhalation is the primary exposure route of concern related to mold for workers, homeowners, and building occupants. When administrative and engineering controls are not adequate to eliminate airborne exposure to mold (or dust containing mold), respirators provide additional protection from inhalation of airborne mold, contaminated dust, and other particulates that are released during dust-generating processes (e.g., remediation work or debris removal).

Respirators provide varying levels of protection. Selecting a respirator to minimize exposure to molds should be based on a qualitative assessment because quantitative data on mold-contaminated environments are not informative. All decisions about respirator selection should be made with knowledge of the relative protective capabilities and the advantages and disadvantages of different respirators.

Standard surgical or dust masks are intended for use only as barriers against large particles and do not provide protection against many airborne particles. Respirators used to protect persons from airborne contaminants (including mold and mold spores) must be certified by CDC's NIOSH. In addition, as specified by the OSHA respiratory protection standard, workers whose employers require them to use respirators must be properly trained, have medical clearance, and be properly fit-tested before they use the respirator. If a worker must use respirators, the worker's employer must develop and implement a written respiratory protection program with worksite-specific procedures and elements. Additional information on respiratory protection is available from OSHA.

PPE Guidelines for Workers in Mold-Contaminated Areas

Exposure to some level of airborne mold is inevitable because molds are found indoors and outdoors. However, demolishing or cleaning heavily mold-contaminated materials outdoors can lead to excessive exposure to mold. The level of exposure to mold outdoors is primarily based on the amount of mold-contaminated material, the amount of mold in the material, and the type of work being performed. The need for PPE (including respiratory, skin, and eye protection) for outdoor workers requires ongoing professional assessment that considers the potential for exposure to mold and the potential for exposure to other hazardous substances that might be in the outdoor work area.

Guidelines summarized below are based on guidelines from OSHA, EPA, and the New York City Department of Health and Mental Hygiene. These guidelines recommend particular respirators on the basis of the size of the area of mold contamination. However, the size criteria are based on general professional judgment and practicality because data are limited related to the extent of contamination to the frequency or severity of health effects.

When determining the potential for airborne exposure to mold and the need for PPE, the size of the area is not the only criterion to be considered. The activities being performed in relation to the mold-contaminated materials are at least as important. Therefore, ongoing professional judgment must always play a part in decisions concerning PPE. For example, any remediation or other work that disturbs mold and causes mold spores to become airborne increases the degree of respiratory exposure. Actions that tend to disperse mold include breaking apart moldy porous materials such as wallboard; destructive invasive procedures to examine or remediate mold growth in a wall cavity; removal of contaminated wallpaper by stripping or peeling; and using fans to dry items or ventilate areas. In addition, health status and other characteristics of the persons potentially exposed to mold also might need to be considered.

The guidelines listed in Box 6.2 should be followed according to professional judgment. For example, more protective

---

Box 6.2   Guidelines for Making Judgments about Mold

CATEGORY I PROTECTION

- Respiratory protection (e.g., N95 disposable respirator); respirators must be used in accordance with the OSHA respiratory protection standard
- Gloves and eye protection

For use while cleaning the following:

- Small isolated areas (≤10 ft^2) of heating, ventilation, and HVAC systems (includes pipes, ducts, and vents)
- Isolated areas (≤100 ft^2) of building materials (e.g., ceiling tiles, small areas on walls, and individual or multiple wallboard panels)

CATEGORY II PROTECTION

- Respiratory protection with full face-piece respirators, with N100, R100, P100 (or for powered air purifying respirators, HEPA) particulate filters; respirators must be used in accordance with the OSHA respiratory protection standard
- Disposable protective clothing covering entire body including both head and shoes
- Gloves

For use while cleaning the following:

- Large contaminated areas (>10 ft^2) of HVAC systems.
- Extensively contaminated (>100 contiguous ft^2) building materials
- Any size area where substantial dust is generated during cleaning or debris removal (e.g., when abrasives must be used to clean contaminated surfaces or plaster walls are being demolished)
- Areas where the visible concentration of mold is heavy (blanket coverage rather than patchy)

---

respirators might be required if toxic contaminants such as asbestos or lead are encountered during cleanup. All workers dealing with large areas of contamination should be properly trained to handle hazardous materials.

PPE Guidelines for the Public (Nonworkers) in Residences and Nonoccupational Settings

The activities (and possible exposure to mold) of persons reentering their homes or working outside might be similar to those of workers. Preventing the creation of dust and limiting exposure to dust are the best ways to minimize exposure to mold. For example, using wet mops or vacuums with HEPA filters instead of dry sweeping dust and debris will decrease the amount of dust in the air.

If building occupants, residents, or anyone must be around mold-contaminated dust, respirators will offer some protection. Particulate respirators (such as NIOSH-certified N95 respirators) can be purchased in safety supply stores and in most home improvement stores. Several factors are required for respirators to provide protection from inhalation hazards:

- The respirator must fit well and be worn correctly. The manufacturer's instructions on the package should be followed. Because respirators are meant to be used by healthy workers who have had training, medical evaluations, and a proper fitting, the amount of protection provided by a respirator to the general public might be much less.
- No U.S. agency tests and certifies respirators for public use. However, NIOSH tests and certifies respirators for use by workers to protect against workplace hazards. Respirators certified by NIOSH will be labeled "NIOSH Approved" and will have an approval label that identifies the hazard it will protect against. The N95 respirator is approved only for particulates including dust in the air from sweeping, sawing, mold removal, and other activities that cause dust. The N95 respirator is not designed to protect against exposure to vapors or gases (e.g., carbon monoxide) and will not provide protection from them.

A properly worn disposable respirator requires that:

- Metal nose piece, if present, is on the top to adjust the fit to the wearer's nose
- NIOSH label is on the bottom outside of the respirator
- Both respirator retaining straps are in place, and they are securing the respirator to the face (some respirators have only one strap)

PPE Guidelines for the Public Not Involved in Cleanup, Debris Removal, or Similar Activities

Persons not involved in activities that disturb mold-contaminated materials have a lower risk for inhalation exposure relative to persons performing those types of activities. Persons collecting belongings, visually inspecting homes or buildings, or doing basic cleanup for short periods in a previously flooded home or building will not usually need to use a respirator.

Guidelines for the Public Unable to Use PPE or at High Risk from Mold Exposure

The effect of exposure to mold varies widely. Persons who might be affected to a greater extent than the majority of adults who are healthy include:

- Persons with respiratory conditions (e.g., asthma) or allergies
- Persons with weakened immune systems (e.g., patients receiving chemotherapy, organ or bone marrow transplant recipients, or persons with human immunodeficiency virus infection or autoimmune diseases)

Persons with special health concerns should consult their healthcare provider if they are concerned about mold exposure.

Symptoms that might seem related to mold exposure might have other causes, such as bacterial or viral infections or other allergies. The level of risk associated with exposure activities and the potential benefit of recommended PPE are unknown for pregnant women, persons older than 65, and children younger than 12; exposure-reducing behavior and respiratory protection might be difficult for children younger than 12.

Using respirators or other PPE might increase health risks for persons with underlying health conditions. Those who have trouble breathing while using a respirator should stop working and contact a doctor or other medical provider. For persons at potentially increased health risk from exposure to mold, persons of unknown or uncertain risk, or persons unable to use respirators, caution is recommended when entering heavily mold contaminated environments, particularly when mold cleanup is occurring. Persons in these categories should avoid such situations if possible.

▶ POTENTIAL HEALTH EFFECTS OF FUNGAL CONTAMINATION

In recent years, the health effects of exposure to mold in built environments have been a subject of intense public concern. These concerns and how they are approached will have important implications for the reconstruction and rehabilitation of cities in states affected by major hurricanes or floods.

Many clinical conditions could be caused by the fungal contamination associated with flooding after major hurricanes or floods. Predicting what might occur is speculative. However, many of these conditions are uncommon and will be recognized only if there is a high clinical index of suspicion. Anticipating what medical problems could be associated with post-flood fungal contamination might help in preventing them by identifying susceptible populations and making recommendations for reducing potentially harmful exposures.

Although the focus of this chapter is on the potential health effects of fungal contamination, other exposures are also of concern. For example, dampness favors the proliferation of dust mites and microorganisms such as bacteria and nontuberculous mycobacteria. Endotoxins (components of the cell walls of Gram-negative bacteria) have strong inflammatory properties. Moisture also can release chemical constituents from building materials. Standing water supports rodent and cockroach infestations and proliferation of mosquitoes. Fecal contamination of the environment raises concerns about protozoal and helminthic parasites. Fungi are not the sole potential cause of many conditions discussed here, and these conditions are only a subset of the conditions of concern to clinicians and public health professionals dealing with the aftermath of major hurricanes or floods.

Overview of Fungal-Induced Diseases

Fungi can cause a variety of infectious and noninfectious conditions. Several basic mechanisms can underlie these conditions, including immunologic (e.g., IgE-mediated allergic), infectious, and toxic. Several of these mechanisms contribute to pathogenesis of a fungal-induced disease. The types and severity of symptoms and diseases related to mold exposure depend in part on the extent of the mold present, the extent of the person's exposure, and the susceptibility of the person (e.g., persons who have allergic conditions or who are immunosuppressed are more susceptible than those without such conditions). Molds produce a variety of volatile organic compounds, the most common being ethanol, which are responsible for the musty odors associated with fungal growth. Exposure to moldy indoor environments is also associated with a variety of upper and lower respiratory tract symptoms.

Institute of Medicine Report on Damp Indoor Spaces and Health

In recent years, the issue of how damp indoor spaces and mold contamination affect human health has been highly controversial. In response, CDC commissioned the Institute of Medicine (IOM) to perform a comprehensive review of the scientific literature in this area. The resulting report was published in 2004 and remains the most current and authoritative source of information on this subject. The IOM listed its findings into four categories:

1. Sufficient evidence of a causal relation
2. Sufficient evidence of an association
3. Limited or suggestive evidence of an association
4. Inadequate or insufficient evidence to determine whether an association exists

"Inadequate or insufficient evidence to determine whether an association exists" does not rule out the possibility of an association. Rather, it indicates that no studies examined the relation or that published study results were of insufficient quality, consistency, or statistical power to permit a conclusion about an association. No conditions exist for which the IOM found sufficient evidence of a causal relation with mold or with damp indoor spaces. Several of the conditions are of particular interest to those engaged in the response to major hurricanes or floods. Sufficient evidence links upper respiratory tract symptoms (e.g., nasal congestion, sneezing, runny or itchy nose, and throat irritation) to damp indoor environments and mold (with exposure to mold often determined by self-report).

Similarly, sufficient evidence exists that there is most likely a link with lower respiratory tract symptoms of cough and wheeze. Sufficient evidence also was found to show that there is a link between damp indoor environments, mold, and asthma symptoms in sensitized persons. In addition, evidence is sufficient to indicate an association between mold exposure and hypersensitivity pneumonitis (HP) in a small proportion of susceptible persons, invasive respiratory and other fungal infections in severely immunocompromised persons, and fungal colonization of the respiratory tract or an infection in persons who have chronic pulmonary disorders.

IgE-Mediated Diseases Caused by Fungi

IgE-mediated, or allergic, responses underlie the most common types of diseases associated with exposure to fungi. Atopy, or the genetic predisposition to form IgE responses to aeroallergens, is an important risk factor. Clinical conditions associated with allergies include allergic rhinitis and asthma. Allergic rhinitis is often associated with allergic conjunctivitis and sinusitis.

Symptoms of allergic rhinitis include sneezing; itching of the nose, eyes, mouth, or throat; nasal stuffiness; clear rhinorrhea; and, if associated with allergic conjunctivitis, red, itchy eyes. If associated with sinusitis, persons also might complain of sinus fullness or postnasal drip, often purulent. Signs on physical examination include pale, boggy nasal mucosa; nasal obstruction; and conjunctival redness. Examination of nasal scrapings or secretions indicates eosinophilic inflammation. If appropriate allergy prick skin testing reagents or in vitro tests for serum specific IgE are conducted, they demonstrate specific IgE-sensitization to causative allergens.

Skin testing reagents and blood tests for the documentation of IgE-sensitization to molds are, with few exceptions, poorly standardized and of unclear sensitivity and specificity. The conventional hierarchy of treatment is avoidance of exposure to inciting agents; pharmacotherapy with antihistamines, decongestants, or anti-inflammatory agents (e.g., nasal steroid sprays); or, as a last resort, allergen immunotherapy. Immunotherapy with fungal allergenic extracts is, with a few exceptions, of unknown efficacy.

Asthma is a disease characterized by episodic, reversible airways obstruction and eosinophilic airways inflammation. Over time, chronic asthma can lead to airways remodeling and irreversible airways obstruction. Persons with asthma often have symptoms such as chest tightness, wheezing, dyspnea, or cough. Physical examination during active asthma might indicate wheezing, but results of examinations between attacks are most often normal. If performed during an active asthma attack, spirometry most often indicates obstruction that reverses with inhalation of a bronchodilator.

Persons with asthma generally exhibit bronchial hyperreactivity to methacholine challenge. However, a small proportion of persons without asthma and a substantial proportion of persons with airway disorders, including chronic obstructive pulmonary disease (COPD), also might exhibit hyperreactivity to inhaled methacholine; therefore, test results must be considered together with other clinical information. Approaches to demonstrating specific IgE sensitization to molds and limitations of available methods are as described for allergic rhinitis. Asthma is associated with airways inflammation that can be demonstrated by examining induced sputum for eosinophils or measuring exhaled nitric oxide, but these tests are often not performed in clinical settings.

Comprehensive guidelines for the staging and treatment of asthma are provided by the National Institutes of Health. Identifying and avoiding triggers, including occupational triggers, is a critical element of treatment. It is important to identify persons with asthma triggered by materials in flood-damaged areas so avoidance measures can be taken. Drug treatment of asthma consists of symptom controllers such as bronchodilators and anti-inflammatory agents (e.g., corticosteroids or leukotriene antagonists). The role of allergen immunotherapy with most fungal agents in treatment of asthma is unclear.

Therapy with monoclonal anti-IgE is a recently developed treatment option that can be used in carefully selected patients when other, less expensive modalities fail to reduce dependence on systemic corticosteroids. The exacerbation of symptoms of asthma is consistently associated with damp buildings. If persons with asthma must engage in activities within damp or mold-contaminated buildings, their asthma should be well controlled before entering these buildings, and those around them should be aware of the signs of asthma symptoms. The onset of symptoms while in damp, moldy environments, especially while wearing PPE, should be an indication to leave the area and to seek appropriate medical care.

Allergic Diseases Associated with Airways Colonization

Allergic bronchopulmonary aspergillosis (ABPA) is a disease that can occur when the airways of persons with obstructive pulmonary diseases (e.g., asthma or cystic fibrosis) become colonized with *Aspergillus fumigatus* or other *Aspergillus* species. Inflammatory responses lead to additional airways damage. Marked worsening of existing asthma is a typical presentation of ABPA. Symptoms include recurrent episodes of bronchial obstruction, fever, malaise, expectoration of brownish plugs, peripheral blood eosinophilia, hemoptysis, and sometimes asymptomatic pulmonary consolidation.

Other features include immediate skin test reactivity to *Aspergillus* spp. antigens, precipitating serum antibodies to *A. fumigatus*, markedly elevated serum total IgE, fleeting lung infiltrates, and central bronchiectasis. Criteria for diagnosis have been published. Airways colonization with other fungal species can result in a similar clinical picture. Although no known relation exists between levels of exposure to *Aspergillus* spp. and development of ABPA, clinicians should suspect and evaluate for the condition when appropriate.

Allergic fungal sinusitis (AFS) is typically noninvasive and it occurs in allergic, immunocompetent patients: most have asthma, and 85% have nasal polyps. Invasive fungal sinusitis can occur in patients who are immunocompromised with illnesses such as diabetes, hematologic malignancies, or immunosuppressive treatments or chronic steroid therapy. Fungal colonization is associated with a characteristic allergic mucin containing high levels of eosinophils. The mere presence of fungi in the nasal passages is not indicative of an active infection.

Hypersensitivity Pneumonitis

Hypersensitivity pneumonitis (HP), also known as extrinsic allergic alveolitis, is a granulomatous interstitial lung disease. A wide range of materials, including fungi, can be inhaled and thus sensitize susceptible persons by inducing both antibody and cell-mediated immune responses. Reexposure of sensitized persons leads to lung inflammation and disease. Building-related HP caused by fungi and bacteria has been well documented. Usually, only a small fraction of those with a given exposure develop HP; therefore, poorly understood host factors play an important role in disease pathogenesis.

The presentation of HP is complex and can be either acute, subacute, or chronic. The acute form is often associated with heavy exposures and characterized by chills, fever, malaise, cough, and dyspnea appearing 4 to 8 hours after exposure and is often confused with pneumonia. The chronic form is thought to be induced by continuous low-level exposure. Onset generally occurs without chills, fever, or malaise and is characterized by progressive shortness of breath with weight loss. Chronic HP can be confused with idiopathic pulmonary fibrosis or other forms of interstitial lung disease.

The diagnosis of HP, especially the chronic form or when presentation is mild, is often missed early in the course of the disease. If it does occur in the aftermath of major hurricanes or floods, a high degree of clinical suspicion is required for detection. In general, when HP is suspected, a clinical and exposure history should be performed. Patients should be asked about their possible exposure to damp and water-damaged areas, farms, birds, hot tubs, and other environments that might cause HP. Environmental sampling for the presence of microorganisms known to cause HP and serologic testing for circulating precipitins can help to establish causative exposures.

Chest imaging using chest radiographs or high-resolution computed tomography scanning of the thorax, lung-function tests, broncholaveolar lavage, and lung biopsy all have roles in diagnosis. Although established criteria exist for the diagnosis of hypersensitivity pneumonitis, in the setting of a documented post-disaster HP outbreak, a noninvasive approach to identifying cases could be more appropriate and cost-effective than requiring conventional diagnostic testing. A recent, large multicenter study indicated that under conditions of low or high prevalence, the following six predictors could be used in combination for noninvasive diagnosis of HP.

- Exposure to a known offending antigen
- Positive precipitating antibodies to the offending antigen
- Recurrent episodes of symptoms
- Inspiratory crackles on physical examination
- Symptoms occurring 4 to 8 hours after exposure
- Weight loss

Optimal treatment is elimination of causative exposures. The IOM report provides information about management of building-related HP that is relevant to reoccupation of structures contaminated by fungi after major hurricanes or floods. Such management includes giving standard medical therapy (e.g., systemic corticosteroids and removing sources of fungal contamination from the environment). In some cases, if efforts to remove mold from a building are unsuccessful in relieving symptoms, the patient might need to move to another home or office.

Inhalation Fevers

Inhalation fever is a general name given to a variety of influenza-like, self-limited syndromes that might be caused by a variety of stimuli. Two types of inhalation fevers are of potential concern after major hurricanes or floods.

Humidifier fever is characterized by fever, respiratory symptoms, and fatigue with onset within hours after exposure to contaminated humidification systems. Obtaining a supportive history is critical to diagnosis. Thermophilic actinomycetes; other bacteria, including species of *Legionella* and *Pseudomonas*; and protozoa have been associated with humidifier fever. Aerosolized endotoxin derived from Gram-negative bacteria has an important role in this condition. Although humidifier fever can be confused with acute HP, it is a short-term ailment and removal from exposure is effective treatment. Humidifier fever is thought to represent a nonspecific inflammatory response to exposure.

Organic dust toxic syndrome (ODTS) has been reported among workers in a variety of agricultural and industrial settings and is thought to involve inhalation exposure to materials with heavy microbial contamination. Etiologic exposures that cause ODTS are often a poorly defined mixture of substances, including fungi, bacteria, and microbial constituents such as endotoxin. ODTS is characterized by fever and influenza-like symptoms, including general weakness, headache, chills, body aches, and cough occurring 4 to 12 hours after heavy exposure to organic dust.

Dyspnea is also sometimes present. Results of chest auscultation and chest radiographs are usually normal. The peripheral white blood count is often elevated during attacks. Accurate patient history is critical for making a correct diagnosis. Although the symptoms resemble those of acute HP, they are not caused by response of the immune system to a specific antigen in the environment. ODTS poses a risk for workers performing renovation work on building materials and is a realistic concern for workers handling heavily contaminated materials in the aftermath of major hurricanes or floods. ODTS is best prevented by minimizing exposure through engineering controls, administrative controls, and respirators. For agricultural workers handling organic dusts, CDC recommends using the most practical respirator with the highest assigned protection factor.

Toxic Effects of Fungi

Certain common molds can produce metabolites with a wide range of toxic activities such as antibiotic (e.g., penicillium), immune-suppressive (e.g., cyclosporine), carcinogenic (e.g., aflatoxins), emetic, and hallucinogenic (e.g., ergot alkaloids). Mycotoxins are fungal metabolites that poison humans and animals. Although ingestion is the most common route of exposure, inhalation and dermal contact also are exposures of concern. Mycotoxin production is dependent not only on species and strain of mold, but also on environmental conditions (e.g., temperature, water activity, light) and growth substrate. Thus, the presence of toxin-producing mold species does not necessarily indicate whether mycotoxins are present.

Mycotoxins were prematurely proposed as the cause of a disease outbreak of eight cases of acute pulmonary hemorrhage/hemosiderosis in infants in Cleveland, Ohio, in 1993 and 1994. The cluster was attributed to exposure to mycotoxins produced by *Stachybotrys chartarum*. Subsequent reviews of the evidence concluded that insufficient information existed and no such association was proven.

Almost all of the known effects of mycotoxin exposures are attributable to ingestion of contaminated food. Health effects from inhalational exposures to toxins are not well documented. IOM found inadequate or insufficient evidence for a link between exposure to damp indoor environments and molds with a variety of conditions that have been attributed to toxicity. Certain case studies of agricultural and remediation workers have described adverse health effects such as skin irritation, skin necrosis, cough, rhinitis, and bloody nasal exudate after inhaling or touching materials with heavy fungal contamination. Whether these effects resulted from exposure to mycotoxins or from a general overload of organic material is unknown. No commercial clinical diagnostic tools are available to determine whether a person's health effect is related to exposure to mycotoxins. Because of the lack of information about noningestion mycotoxin exposure and adverse health effects in humans, precautions should be taken when handling heavily contaminated building materials.

Fungal Infections

No reports of increased fungal infections related to floods in the United States exist. However, anecdotal case reports of fungal infection after floods include *Apophysomyces elegans* wound infection in a man who sustained traumatic injuries after the southeast Asian tsunami in December 2004. *A. elegans* belongs to the Zygomycetes class of fungi. Infections are most commonly seen in immunocompromised and diabetic patients, and rarely in immunocompetent persons. The cause of infection in immunocompetent persons is usually cutaneous trauma with direct implantation of fungal organisms into the wound from soil contamination.

Theoretically, infection with fungal species that contaminate buildings, building constituents, and the environment after major hurricanes or floods is a potential concern. In general, persons with impaired host defenses (especially if impaired because of cell-mediated immunity or neutropenia) suffer the most severe types of fungal infections. However, invasive fungal infections also can occur in persons with normal host defenses and, in certain situations, can be life threatening. Persons at greatest risk for invasive fungal infection from heavy fungal contamination after major hurricanes or floods are those with impaired host defenses.

Any impairment in cell-mediated immunity or neutropenia (e.g., HIV infection, leukemia, lymphoma, and diabetes mellitus) increases risk for many types of invasive fungal infections. Severely immunosuppressed persons, such as solid-organ or stem-cell transplant recipients or those receiving cancer chemotherapy agents, corticosteroids, or other agents inhibiting immune function, are at much higher risk for locally invasive infections of the lungs, sinuses, or skin and systemic infections. *Aspergillus* spp., zygomycetes, and *Fusarium* spp. are particularly important problems. These serious infections are often fatal, even with aggressive antifungal therapy.

Protective measures, such as HEPA filtration, implemented during periods of extreme susceptibility to invasive fungal infections are well established and effective in hospitals. However, preventive measures outside the hospital are less well established. Current guidelines emphasize the importance of avoiding areas of high dust (e.g., excavation sites, building construction or renovation sites, chicken coops, and caves) and factors associated with fungal infections (e.g., foods that increase a person's risk for fungal exposure).

Obstructive pulmonary diseases such as asthma, cystic fibrosis, and COPD, might predispose persons to airway colonization with *Aspergillus* spp. Inflammatory host responses to colonization can lead to ABPA. *Aspergillus* spp. also can cause invasive or semi-invasive infection in persons with COPD, especially in those being treated with corticosteroids. Chronic necrotizing pulmonary aspergillosis is an indolent condition observed in persons with underlying lung disease.

Colonization of lung cavities (e.g., tuberculosis cavities or emphysematous blebs) by *Aspergillus* spp. can cause pulmonary aspergillomas (fungus balls), which are conglomerations of *Aspergillus* spp. hyphae matted together with fibrin, mucus, and cellular debris. These often do not cause symptoms, but they can be associated with hemoptysis. An exposure-response relation has never been established linking levels of exposure to *Aspergillus* spp. with development of any of these conditions. Therefore, to what degree exposure to fungal contamination after major hurricanes or floods would increase any risk is unclear. However, despite unknown benefit, persons with clinically significant obstructive pulmonary diseases (e.g., asthma, cystic fibrosis, COPD), and persons with cavitary lung disease from conditions such as tuberculosis should avoid airborne exposure to materials that have become heavily contaminated with fungal growth in the wake of major hurricanes or floods.

Persons with normal host defenses are also subject to fungal infections, and persons with impaired host defenses can acquire any of these, often with greater severity. Ocular, skin, and superficial infections occur in those with normal host defenses and range from the relatively common (e.g., ringworm, athlete's foot) to the relatively rare (e.g., sporotrichosis). Of particular relevance in areas with fungal contamination after major hurricanes or floods are organisms that cause localized skin and superficial infections following traumatic inoculation with soil and plant saprophytes, which are found in air, soil, and plant materials. For example, *Scedosporium apiospermum* (*Pseudallescheria boydii*) can be recovered from polluted water, sewage, swamps, and poultry or cattle manure.

Although rare in the United States, this organism can cause a soft tissue infection called Madura foot, a mycetoma in which the draining sinuses show white grains containing fungal elements. This organism can also produce septic arthritis or osteomyelitis after penetrating trauma. *Sporothrix schenckii* is a dimorphic fungus that produces soft tissue infections after traumatic inoculation from a contaminated environmental source, such as sphagnum moss, roses, plant seedlings, and other vegetation. Lymphocutaneous lesions are the hallmark of sporotrichosis, as the organisms spread through the local lymphatics after primary inoculation. A high degree of clinical suspicion is needed to diagnose the less common, locally invasive fungal infections. Diagnosis is made by histopathology and culture after biopsy of the affected lesion. Histopathology must be performed to verify that a recovered isolate is the cause of disease and not an environmental contaminant. Culture must be performed to identify the agent correctly. Fungal isolates are identified in a clinical mycology laboratory.

Exposures that result in invasive pulmonary mycoses in persons with normal host defenses are generally thought to occur outdoors where active disturbance of a reservoir has occurred. The mode of transmission is inhalation of fungal spores. Person-to-person transmission of pulmonary mycoses does not occur. Diseases relevant to flood prone areas such as the Gulf Coast states include histoplasmosis and blastomycosis. Histoplasmosis is unlikely to be increased as a result of fungal contamination after major hurricanes or floods. The condition is caused by *Histoplasma capsulatum*, a dimorphic fungus found in soil enriched with the droppings of birds and bats. Areas with endemic disease in the United States include the Mississippi and Ohio River valleys, but cases have occurred in other parts of the United States. Many persons develop no symptoms when exposed to *H. capsulatum* in an endemic setting.

Blastomycosis is a potential problem after major hurricanes or floods in areas with endemic disease because it can cause serious disease even in those with normal host defenses. Blastomycosis is caused by the dimorphic fungus *Blastomyces dermatitidis*. The organism is found in moist soil, frequently along streams or rivers enriched with decaying vegetation. In the United States, the organism is most commonly found in states surrounding the Mississippi and Ohio rivers. An area in Louisiana about 70 miles from New Orleans has endemic blastomycosis. In Louisiana, cases occur at an incidence of about 1 to 10 per year, mostly in the area of Washington Parish where the condition is endemic. Outbreaks have been associated with manipulation of decaying vegetation or recreational activity near lakes or rivers. The incubation period is not certain but, on the basis of data from outbreaks, appears to be about 45 days, ranging from weeks to months.

The clinical spectrum of blastomycosis includes asymptomatic infection, acute or chronic pneumonia, and disseminated disease. Pulmonary infection can mimic acute bacterial pneumonia or tuberculosis with progression to acute respiratory distress syndrome. Alveolar infiltrates, mass lesions that mimic bronchogenic carcinoma, and fibronodular interstitial infiltrates are the most common radiographic findings. Disseminated blastomycosis often appears as ulcerative skin lesions with multiple necrotic bone lesions in the vertebrae, skull, or long bones.

Culture of lesions or histopathologic evidence from infected tissue is required for diagnosis of blastomycosis. Direct microscopy of pus, scrapings from skin lesions, or sputum showing thick-walled broad-based budding yeast cells that are 5 to 15 μm in diameter supports a presumptive diagnosis of blastomycosis and might, in the appropriate clinical setting, prompt the initiation of antifungal therapy. Serologic tests can be performed on serum from patients showing signs of suspected pulmonary blastomycosis or with suggestive skin lesions. A positive immunodiffusion (ID) test, showing a precipitin band with the *Blastomyces* A antigen, is highly specific for the disease and does not require paired serum samples. However, the sensitivity is poor (33–88%), so a negative ID test does not rule out the disease. For cases with negative results, the test should be repeated in 3 to 4 weeks after the initial sampling. The complement fixation (CF) test for blastomycosis has poor sensitivity and specificity.

Fungal brain abscesses are uncommon in healthy persons. The primary infection results from inhalation of infectious conidia from the environment; the route of infection appears to be hematogenous dissemination from the lungs. Of particular interest after major hurricanes or floods is *S. apiospermum* (*P. boydii*). Many case reports document patients with focal neurologic defects caused by multiple brain abscesses weeks or months after nearly drowning. The organism apparently spreads hematogenously after initial aspiration of sewage-laden water (from floods, lagoons, or bayous) into the lungs. Near drowning presumptively results in a massive inoculation of mold into the lungs.

▶ PREVENTING FUNGAL CONTAMINATION AFTER HURRICANES OR FLOODS

Persons should reduce their exposure to molds as much as possible (with the realization that fungi are ubiquitous). Persons with underlying or induced immune suppressed conditions or diseases caused by immune sensitization to fungal constituents present in mold growth should be especially careful to reduce exposure. If exposure to heavily mold-contaminated materials is unavoidable, persons should use appropriate administrative, engineering, and personal protection controls. Because a person's likelihood of developing adverse health effects from mold exposure depends on the type of exposure and on individual susceptibility, precautionary measures need to be customized. Recommended measures are based on professional judgment because of lack of available scientific evidence. For example, no research studies have evaluated the effectiveness of personal protective equipment in preventing illness from mold exposure.

Total avoidance of heavily contaminated buildings or other high exposure situations is suggested for persons with specific underlying conditions such as profound immune suppression. Respiratory protection, dermal protection, and occlusive eye protection recommendations are customized to various populations and exposure-associated activities. Repeated or prolonged exposure probably poses a greater health risk than do exposures of a similar intensity, but short duration. Preventive precautions are especially important for persons who expect to be highly exposed for a long time.

Public Health Strategies and Recommendations for State and Local Officials

Recommendations from CDC are for protecting and monitoring the health and safety of workers and residents who enter, repair, or destroy flooded buildings. The recommendations are focused on limiting human exposure to mold and other microbial agents and preventing any adverse health effects related to such exposure. Several factors are assumed:

- In the aftermath of major hurricanes and floods, buildings or materials soaked for more than 48 hours are contaminated with mold unless proven otherwise by inspection or adequate environmental sampling or cleaned according to the EPA's recommendations.
- Workers and residents might be exposed to high levels of mold-related contaminants.
- Sufficient evidence exists of an association between adverse health outcomes and exposure to damp indoor environments or materials contaminated with fungal growth.

- Insufficient evidence exists for establishing health-related guidelines on the basis of concentrations of mold (quantitative measure) or species of mold (qualitative measure) in either indoor or outdoor environments.
- Allergen testing to determine the presence of IgE to specific fungi might be a useful component in the complete clinical evaluation and diagnosis of mold-related allergies and in the decision to avoid exposure to fungal allergens that might be causing allergic symptoms. However, testing for IgE sensitization to molds has important limitations. Allergens used in these tests are often poorly standardized and the tests often have unclear sensitivity and specificity. In addition, allergen testing is not relevant to diseases that are not mediated by IgE.
- Clear, concise, and practical recommendations and actions are necessary to limit exposure to mold and to prevent mold-related health outcomes where possible.

Assessing Exposure to Mold

Exposure assessment is usually a critical step in determining whether persons are exposed to a hazard at a level that could have an adverse health effect. The mere presence of a chemical or biologic hazard in the environment is insufficient to create a public health hazard. The contaminant must be present in an environmental medium (e.g., air, water, food, and dust) that allows it to come in contact with persons and move along a biologic pathway (e.g., inhalation, ingestion, and absorption). In addition, the concentration of the contaminant must be sufficient to create a biologic response that leads to an adverse health outcome.

Mold and its spores exist in damp materials. Disturbing mold releases potentially hazardous particulates into the air, which can then be drawn into the sinuses and lungs. Although molds might also directly attack the skin or openings in the skin, the most common route of exposure is through the air and into the body by inhalation. Environmental sampling for molds has limited value and, in most instances, is not needed after major hurricanes or floods. Building interiors should be assumed to be substantially contaminated with mold in the following circumstances:

- The building was saturated with water for more than 48 hours.
- Visible mold growth is extensive and in excess of that present before a major hurricane or flood.
- Signs of water damage are visible or mildew odors are strong.

Exposure to materials and structures contaminated with mold should be assumed to present a potential health risk regardless of the type of mold. Risk for illness does not necessarily vary with the type of mold or the extent of contamination.

Preventing Excessive Exposure to Mold

Preventing excessive exposure to mold is the best way to avoid harmful health consequences. The preferred approach to preventing mold exposure is to prevent water from infiltrating a building or damaging household goods and structures. After major hurricanes or floods, substantial water damage and mold growth might occur in many buildings.

If left undisturbed, mold is generally not a hazard, and most persons will not be adversely affected by moderate exposure to mold. However, in the aftermath of a major hurricane or flood, remediation activities within buildings will disturb any mold that is growing and lead to exposure. To prevent excessive exposure to mold in contaminated areas that are disturbed, persons who enter those areas should implement environmental controls (e.g., suppression of dust and isolation of the contaminated area), use PPE, or both.

Preventing human exposure to mold and health effects from such exposure depends on the following three factors.

- The likely concentration of mold in or on the building fabric or materials.
- The duration and type of activity undertaken in the mold-contaminated area.
- The susceptibility of the person entering the area to the various health effects.

Four methods for preventing exposure to mold can be used in combination:

- Avoiding areas thought to be mold-contaminated
- Using environmental controls
- Using PPE
- Employing strict personal hygiene

Avoidance

The following persons should avoid mold-contaminated environments entirely.

- Transplant recipients, including those who received organ or hematopoietic stem cells during the preceding 6 months or who are undergoing periods of substantial immune suppression

- Persons with neutropenia (neutrophil count <500/μL) attributed to any cause, including neoplasm, cancer chemotherapy, or other immunosuppressive therapy
- Persons with CD4+ lymphocyte counts less than 200/μL attributed to any cause, including HIV infection
- Other persons considered by their physicians to have profoundly impaired antifungal host defenses caused by congenital or acquired immunodeficiency

The following persons might be able to tolerate limited exposure, but they should consult with their physicians and should consider avoiding areas where moldy materials are disturbed.

- Persons receiving chemotherapy for cancer, corticosteroid therapy, or other immunosuppressive drug therapy, as long as neutropenia or CD4+ lymphopenia are not present
- Persons with immunosuppressive diseases such as leukemia, lymphoma, or HIV infection, as long as there is not marked impairment in immune function
- Pregnant women
- Persons over 65 years of age
- Children less than 12 years of age, particularly infants,
- Persons with chronic, obstructive, or allergic lung diseases

All buildings with extensive mold contamination require remediation before rehabilitation. Remediation includes structural repairs to prevent additional water intrusion, removal of mold-contaminated materials that cannot be adequately cleaned and decontaminated, and cleaning and decontamination of mold-contaminated materials that can withstand such treatment. Health-care facilities and other locations that house highly susceptible persons require special attention. These facilities must be adequately remediated before being occupied by highly susceptible persons. Guidelines for remediating health-care facilities include:

- "Remediation and Infection Control Considerations for Reopening Healthcare Facilities Closed due to Extensive Water and Wind Damage"
- "Checklist for Infection Control Concerns when Reopening Healthcare Facilities Closed due to Extensive Water and Wind Damage"

Use of Environmental Controls

Examples of environmental controls include isolation or containment of the contaminated area, ventilation of the area, and suppression of dust in the area (e.g., by wet-mopping the mold-contaminated surfaces to reduce airborne mold concentrations). Certain methods of isolation can be used to minimize mold exposure. For example, workers operating heavy equipment during the demolition and removal of mold-contaminated materials can be isolated in sealed, positive-pressure, air-conditioned cabs that contain filtered air recirculation units. Another method of isolation is sealing off of mold-remediation areas in occupied, mold-contaminated buildings. However, such isolated areas must also be adequately ventilated.

Preventing the creation of dust and limiting exposure to dust are essential to minimizing exposure to mold. When cleaning up dust, workers should use wet mops or vacuums with HEPA filters instead of dry sweeping.

Respirators

Inhalation is the primary exposure route of concern related to mold for workers, homeowners, and building occupants. Environmental controls are sometimes inadequate to control airborne exposure to mold or dust containing mold. In such cases, respirators protect persons from inhaling airborne contaminated dust and other particulates released during dust-generating processes (e.g., cleanup or debris removal). Recommendations on when to wear a respirator depend on the severity of mold contamination, whether the person's activity is such that mold or particles containing endotoxin or other microbial agents are likely to be released into the air, and the person's health status.

The following recommendations are made with the assumption that extensive mold contamination is present.

Respiratory Protection

Recommendations for use of respirators include:

- Healthy persons who are in a building for a short time or who are in a place where activity minimally disturbs contaminated material might not need a respirator.
- Persons engaged in activities that moderately disturb contaminated material (e.g., light cleaning by removing mold from

surfaces with a wet mop or cloth) and persons with health conditions that place them at risk for mold-related health problems should use at least an N95 respirator that is certified by NIOSH.

- Persons doing remediation work that involves extensive exposure to mold should have respiratory protection greater than that provided by a NIOSH-certified N95 respirator. Full face-piece respirators that have NIOSH-certified N100, R100, or P100 particulate filters are recommended. For powered air-purifying respirators, a HEPA filter is recommended.
  - Respirator selection is made after considering the characteristics of the work activities; the specific exposures of concern; and the protection factors, advantages, and disadvantages of various respirators.
  - The determination of whether a person will have extensive exposure to mold should be based on several factors, including the size of the mold-contaminated area, the type of mold-contaminated material, and the activities being performed. Guidelines based solely on area of contamination define extensive contamination as being >100 ft^2.
  - Formal fit testing is recommended for anyone engaging in remediation work causing extensive exposure to mold.

Guidelines for respiratory protection use:

- Respirators must fit well and be worn correctly.
- NIOSH tests and certifies respirators for use by workers to protect against workplace hazards. Respirators certified by NIOSH have "NIOSH Approved" written on them and have a label that identifies the hazard the respirators protect against.
- The N95 respirator is approved only as protection against particulates (including dust) and will not protect persons from vapors or gases such as carbon monoxide.

Eye Protection and Protective Clothing

For information about protecting the eyes and use of protective clothing, see Skin and Eye Protection and Protective Clothing sections earlier in this chapter. This CDC/NIOSH publication, "Eye Safety: Emergency Response and Disaster Recovery," provides further information on eye protection.

Hygiene

Disposable PPE should be discarded after it is used. Such equipment should be placed into impermeable bags and usually can be discarded as ordinary construction waste. Appropriate precautions and protective equipment for biocide applicators should be selected on the basis of the product manufacturer's warnings and recommendations (e.g., goggles or face shield, aprons or other protective clothing, gloves, and respiratory protection). Reusable protective clothing should be cleaned according to the manufacturers' recommendations after the product has been exposed to mold. Hands should be washed with clean potable water and soap after gloves are removed.

Health officials should consider whether their agencies should supply PPE to residents who might not otherwise be able to acquire the necessary equipment. Providing PPE to the local population would require substantial resources and a mechanism for distributing them.

Mold-Contaminated Areas and Mold Removal

The following precautions to reduce exposure to mold should be observed during cleanup.

- Items that have soaked up water and that cannot be cleaned and dried should be removed from the area and discarded.
- Dehumidifiers and fans blowing outwards toward open doors and windows can be used to remove moisture.

The procedure to remove mold from hard surfaces that do not soak up water (i.e., nonporous) is as follows.

- Mix 1 cup of bleach in 1 gallon of water.
- Wash the item with the bleach mixture.
- Scrub rough surfaces with a stiff brush.
- Rinse the item with clean water.
- Dry the item, or leave it to dry.

Additional Safety Guidelines for Mold Cleanup

The procedure for cleaning hard surfaces that do not soak up water to prevent mold growth is as follows:

- Wash the surfaces with soap and clean water.
- Disinfect with mixture of 1 cup of bleach in 5 gallons of water.
- Allow to air dry.

Persons cleaning moldy or potentially moldy surfaces should observe the following.

- Wear rubber boots, rubber gloves, and goggles when cleaning with bleach.
- Open windows and doors to get fresh air.
- Never mix bleach and ammonia because the fumes from the mixture can be fatal.

▶ HEALTH-OUTCOME SURVEILLANCE AND FOLLOW-UP

State and local public health agencies do not generally collect information on the conditions related to mold exposure. In situations where there are large numbers of flooded and mold-contaminated buildings, such as occurred in New Orleans after hurricanes *Katrina* and *Rita* in fall of 2005, the repopulation of those once-flooded areas probably will expose a large number of persons to potentially hazardous levels of mold and other microbial agents.

Efforts to determine the health effects of these exposures and the effectiveness of recommendations to prevent these adverse health effects require a surveillance strategy. Developing such a strategy requires that federal and local health agencies work together to monitor trends in the incidence or prevalence of mold-related conditions throughout the recovery period.

Monitoring trends in health outcomes that might be related to mold exposure will require substantial human and financial resources and will face several challenges. Health outcomes that might be attributed to mold exposure fall into several broad categories. Some potential health outcomes are rare, difficult to diagnose, and relatively specific for fungal exposure (e.g., blastomycosis). Other health outcomes are relatively easy to diagnose, but they have numerous etiologic factors and are difficult to attribute specifically to mold exposure (e.g., asthma exacerbations). Tracking different health outcomes that might be caused by mold exposure requires different surveillance methods. In some cases, follow-up research will be needed to verify that surveillance findings and health outcomes are the result of mold exposure. For some conditions, difficulties in interpreting trends and in relating the outcome to mold exposure might suggest that surveillance is not an appropriate public health approach.

Results of surveillance and follow-up activities will help CDC refine the guidelines for exposure avoidance, personal protection, and cleanup. In addition, these activities should assist health departments to identify unrecognized hazards.

Surveillance

Public health agencies should consider collecting health outcome information from health-care facilities to monitor the incidence or prevalence of selected conditions. State or local agencies should determine the feasibility of this approach and consider the required resources available or attainable to accomplish this goal. Institutions from which data could be collected include hospitals, emergency departments; clinics; and, for some outcomes, specific subspecialty providers. Surveillance requires the establishment of case definitions and reporting sources; development of reporting mechanisms; training of data providers; and the collection, analysis, and reporting of data. The surveillance data should be used to identify increases in disease that are substantial enough to trigger public health interventions or follow-up investigations to learn the reason for the increase and establish targeted prevention strategies.

Public health agencies should consider the need for clinicians to report cases of known or suspected mold-associated illnesses (e.g., invasive fungal disease, blastomycosis, hypersensitivity pneumonitis attributed to mold contamination, ODTS attributed to contaminated dust exposure, and alveolar hemorrhage in infants) to public health authorities for tracking and follow-up investigations. Providers caring for patients at high risk for poor health outcomes related to mold exposure could be targeted. For example, hematologists, rheumatologists, and pulmonologists might care for many patients at risk for invasive mold infections because of underlying malignancies and immune suppression. Enhancing provider-based surveillance requires targeting and educating providers; developing reporting mechanisms; and collecting, analyzing, and reporting data.

Public health agencies should consider the need for the establishment of laboratory-based surveillance as an efficient method for monitoring mold-related illnesses that involve laboratory analyses (e.g., invasive fungal disease, blastomycosis, invasive aspergillosis, histoplasmosis, *Aspergillus* preceptins, zygomycosis, and fusariosis).

Clinical Care

Health-care providers should be alert for unusual mold-related diseases that might occur (e.g., hypersensitivity pneumonitis, ODTS, and blastomycosis). Otherwise, such diseases might not be recognized. Scientific evidence is insufficient to support the routine clinical use of immunodiagnostic tests as a primary means of assessing environmental fungal exposure or health effects related to fungal exposure. Health-care providers who care for persons who are concerned about the relation between their symptoms and exposure to fungi are advised to use immunodiagnostic test results with care and only in combination with other clinical information, including history, physical examination, and other laboratory data. If appropriate allergy prick skin testing reagents or in vitro tests for serum specific IgE are available, they can be used to show specific IgE-sensitization to causative allergens.

Unfortunately, skin-testing reagents and blood tests documenting IgE-sensitization to molds are, with few exceptions, poorly standardized and of unclear sensitivity and specificity. The conventional hierarchy of treatment for allergic diseases includes avoiding exposure to inciting agents; pharmacotherapy; and, as a last resort, allergen immunotherapy. Immunotherapy with fungal allergenic extracts is, with a few exceptions, of unknown efficacy. Clinicians should report cases of mold-induced illness to local health authorities to assist in surveillance efforts.

Mold in Schools

7

Mold in schools can be a big problem. Evidence of precisely what is an acceptable level of mold is debatable. If you research the issue you may find multiple opinions. Schools are often talked about within the industry. The fact that children are involved, and potentially at risk, is the reason interest is keen when deciding how to handle mold in schools.

Official regulations for mold control and abatement are not as abundant as they are for other risks, such as lead or asbestos. Still, this is not to say that mold cannot be a serious problem. Not everyone suffers the same when exposed to mold. One person may not show any signs of distress. Another individual may experience severe respiratory problems.

I have worked with mold for many years as a contractor. My daughter had such a struggle with mold in a home where we lived for a short time that it forced us to move to another house. Normally, my daughter doesn't show signs of allergies. But, the mold was making her very ill. After we moved, her condition returned to normal. So, I have first-hand experience with mold as both a contractor and a father of a child who had difficulty breathing due to mold spores under her bedroom floor.

In writing this book I have done extensive research. For this chapter I decided to use a model that the state of Minnesota chose for dealing with mold in its schools. In doing this I am able to show you a good example of local recommendations and requirements while giving you a strong overview of your options when working with schools that are contaminated with mold. Mold contamination in the indoor environment is a complex issue. While scientific understanding about health effects and growth factors is evolving, there is currently considerable scientific uncertainty. Acknowledging this uncertainty, this guidance represents "best practices" advice that is general in nature for the types of situations many Minnesota school officials have struggled with recently.

There are no state or federal laws that specifically mandate how mold must be remediated. There are no consensus standards or laws about how much mold or what kind of mold is acceptable in a school or work place. Nevertheless, mold is a significant public health issue, and the Minnesota Department of Health (MDH) has adopted a health-protective precautionary view that "no amount of visible mold growth is acceptable in occupied spaces." It is prudent for school officials to make a reasonable and good faith effort to address mold problems following the recommendations provided in this document.

MDH considers all molds potentially harmful when they are allowed to grow indoors. Dead or dormant mold may also be harmful when breathed. Mold contamination of the indoor environment (including schools) has been linked to discomfort and health problems including allergy reactions, asthma symptoms, irritant effects, respiratory problems, and a variety of other nonspecific health complaints. In addition to these potential health effects, remediation workers who do not use appropriate personal protection may develop hypersensitivity pneumonitis or organic dust toxic syndrome. The longer mold is allowed to grow indoors, the greater the likelihood it may become airborne and cause adverse health effects. When not corrected effectively, mold problems can spread to previously unaffected areas, which may increase health risks to occupants. Notification of occupants should be considered as noted in Box 7.1.

Box 7.1 Explicit Hazard Warning Requirements

Molds (and other microbial contaminants) present risks to persons working in contaminated environments. Those risks are different from the chemical risks typically addressed by workplace hazard communications. The Legislature might wish to consider enacting a requirement that workers and building occupants be advised explicitly of hazards posed by mold and other microbial contamination affecting work areas. That could be accomplished by requiring that the Hazard Evaluation System and Information Service (HESIS) "Molds in Indoor Workplaces" flier or comparable information be posted in workplaces along with other normal workplace health and safety communications. Alternatively, this need might be met through administrative regulation on the subject of workplace health and safety notices.

Remediation of mold refers to the process of removing contamination coupled with steps to modify the indoor environment to prevent the recurrence of growth. In many cases it is necessary for the remediation process to include engineering controls and other protective measures to prevent or minimize potentially harmful exposures to workers and occupants. The objectives of any mold remediation project are:

- Correct the underlying moisture problem
- Effectively and safely remove fungal contaminated material, including the mold contaminants in settled dust
- Control contaminants during remediation
- Repair property damage and prevent future loss to building materials and contents

While it is best to address the moisture problem first, this may not be possible. In cases where solving the moisture problem must be delayed, it is still prudent to remove the mold. However, the moisture problem must be addressed as soon as possible; otherwise, mold growth will likely return.

Since important details may vary from one situation to another, each mold problem can be uniquely complex (see Box 7.2). The ability to evaluate and respond to the many issues involved can also vary from school to school. Common sense and a considerable degree of judgment are necessary to determine how to best remediate a specific mold problem. Hence, this document

---

Box 7.2 Points to Ponder

There are several points to ponder when assessing mold conditions. School officials should ensure that remediation of contamination is planned and carried out carefully, especially when contamination is, or is suspected to be, extensive. When facing a suspected mold problem, school officials should consider the following.

- Are there immediate or likely health concerns?
- What is the extent of the problem?
- Are building materials structurally or functionally compromised?
- Can the moisture problem be remediated to prevent future mold problems?
- What are the school officials' liability concerns?
- What are the school officials' public relations concerns?
- What are the remediation options?

intentionally allows school officials considerable flexibility in determining how to respond to mold problems within their facilities. The information is applicable to all but the very routine and small mold problems.

▶ COMMUNICATION

When mold issues arise, school officials should expect staff, students, parents, teachers' unions, the media, or possibly the larger community to have a variety of questions. Underlying many such questions are concerns that need to be recognized and addressed. Open communication about health concerns can foster cooperation and early action vital to the efficient and successful resolution of mold problems. Without it, problems can be made worse and solutions delayed by frustration, anxiety, and distrust.

When a mold problem is perceived as a potential health threat (especially to children), it is predictable that people may become distrustful, anxious, and even openly hostile. This is especially true when parents feel that appropriate actions and safeguards are not being taken, that information has been withheld from them, or that their concerns are not being taken seriously. To manage expectations and prevent unnecessary anxiety, it is essential to effectively deliver complete and accurate information to affected stakeholders about the nature of the problem and the school officials' response. School officials should anticipate common questions and respond to all issues promptly and openly.

When mold problems are small and will likely be corrected through routine custodial practices, extensive communication efforts are often not necessary. However, due to the widespread attention recently given to mold, school officials should inform key stakeholders early whenever potential problems are noticed, such as water leaks, mold growth, or unusual complaints that could signal mold contamination. By acknowledging the existence of even simple problems and explaining how they will be handled, school officials can demonstrate their commitment to protecting building occupants. Early, proactive communication can avert rumors and the perception that information has been concealed.

When mold contamination is extensive, health concerns have been raised, or when remediation will disrupt normal school operations, school officials should develop a communication strategy and make sure it is followed. In all such cases it is critical to be open, honest, and direct. All findings regarding the problem should be fully and promptly shared with the community, especially interested members. It is best to create an opportunity for discussion of these findings. MDH has not identified any circumstances where information should be withheld when questions about mold in a school are raised. The following are communication priorities:

1. Demonstrate that occupants' health and safety is of utmost concern and how potential risks are minimized.
2. Supply appropriate details of project goals, findings, and activities.
3. Provide a mechanism for open, ongoing two-way dialogue between school officials and the affected groups or individuals including a means for occupants to share their observations and theories about problems and potential patterns.

Communication methods that some schools have used effectively include meetings (include question-and-answer opportunities), press releases, memoranda, postings, and flyers. The frequency of messages, methods of communication, and degree of formality should be based on the scope of the project and the audiences' needs and interests (if in doubt, over-communicate). If possible, school officials should identify and attempt to include key stakeholders in the communication planning process, such as building staff, teachers, union representatives, students, parents, the medical community, and the media. These persons may help school officials anticipate and more fully understand the community's concerns. They may also help identify ways to reach the appropriate audiences and become a part of the communication network. When respected community members are involved and they help to explain the issues, additional trust and credibility may be gained, enhancing the effectiveness of the messages.

To ensure that information is consistent and accurate, school officials should identify a single point person to whom all requests for information about the remediation project are referred. For example, the district's designated indoor air quality coordinator may be a logical choice since all Minnesota public school districts must have such a coordinator, and this person should be knowledgeable of how the district handles indoor air quality issues.

Remediation plan details should be made available to all affected parties early so that their concerns are understood and may be accommodated before work begins. Once remediation has begun, school officials should continue to provide updates to keep the interested community members abreast of progress and target completion dates.

▶ EVALUATION OF INVESTIGATION FINDINGS

Correcting a mold problem requires understanding the extent of the problem and the underlying causes. In some cases, this is fairly simple, such as when an obvious moisture source has affected only a limited area resulting in easily observable visible mold. However, this can be difficult when the source(s) of moisture, their interaction with building conditions, or the location(s) of the growth are not readily apparent.

When the mold problem is relatively straightforward and can be corrected through routine custodial practices, it should be remediated as promptly as health and safety practices and procedures allow. When a complex mold problem exists, it is wise to carefully assess the problem thoroughly and objectively before beginning remediation. Resist temptation to address only the easiest, most obvious evidence of contamination without looking for hidden growth or reservoirs of contamination. To achieve a durable and effective solution, it is also imperative to understand the reason(s) for the moisture problem(s). Knowing the source of the excess moisture is vital to correcting it and preventing recurrence of the problem. Identifying the pathways the moisture may have taken can help locate hidden mold growth.

The success of remediating a large-scale mold problem ultimately depends on how well the moisture and contamination problem is understood. If planning the remediation relies heavily on reports of past investigations, the accuracy and completeness of those efforts should be objectively assessed. Review the findings of the reports and evaluate how completely the important issues were assessed. Do not assume that past findings necessarily reflect current conditions. Consider whether the conclusions and recommendations are logical in light of all available information, especially any recent findings. Several sections of the MDH guidance document "Recommended Best Practices for

Mold Investigations in Minnesota Schools" may be helpful in critically reviewing investigation methods and reports.

Further investigation needs may be identified if factors that may be critical to the remediation have not been examined adequately. When moisture problems or contamination are extensive, an informed and thorough inspection of the affected and possibly other related building areas and systems might be needed (including elements of the building envelope). An inventory should be made of all visibly moldy surfaces and materials that are water damaged. Consult "Recommended Best Practices for Mold Investigations in Minnesota Schools," or seek assistance from a professional with experience in assessing buildings for mold and moisture problems if further investigation is needed.

▶ DETERMINING REMEDIATION SCOPE

After gaining a reasonable understanding about the extent of mold contamination and the source(s) of excess moisture, school officials should determine the scope of remediation best suited to the problem. MDH recommends a graded approach based on the criteria below and particular characteristics of the problem. School officials should consider site- and situation-specific details regarding the contamination severity and the nature of the underlying moisture problem, potential health and safety concerns related to remediation activities, and the availability of district resources to correct the problems.

Three categories—"minimal," "moderate," and "major"—are used in this document to characterize the complexity of the contamination problem and the potential for exposure of building occupants. These categories are used throughout this document with specific reference to the definitions provided in this section. They are based on quantifiable and nonquantifiable factors; they are not intended to be applied strictly, but to describe and rank the hazards and remediation factors in a relative sense. This deliberate subjectivity underscores the need for school staff to develop remediation practices and procedures that are site- and situation-specific, yet can be adapted if new information arises during the remediation.

It is crucial to remember that the extent of hidden mold growth may be much greater than what is readily visible from within occupied spaces, if the moisture problem originated in or has impacted closed spaces or cavities. In such cases, destructive techniques may be used carefully to access and inspect the inside surfaces of floor, wall, and ceiling cavities. Whenever there is information suggesting that additional contamination may be uncovered during remediation or investigation, increased contaminant control and personal protective measures should be used. Plans should be made flexible to allow for any necessary revision of the project's scope, such as adjusting work practices and procedures if unforeseen contamination or other complications are encountered.

Persons responsible for planning the remediation should review and discuss the three criteria below (amount of mold growth, degree of contamination, and potential for releasing contaminants) to determine which category best describes the problem. Initially, the three criteria should be considered with roughly equal importance.

Amount of Mold Growth

How to best remove mold contamination depends in part on how much is present. For practical reasons, estimating the surface area of visible mold growth and the density of the growth are adequate starting points to approximate and bound the problem. Heavily contaminated materials may contain many more mold particulates than very light and superficial growth over a larger area. Project managers should apply commonsense judgment and consider the amount of the mold growth in conjunction with the other two criteria. The estimated cleanup area should include at least two feet beyond that which is visibly impacted, because mold colonies may extend outward from the visible growth for some distance.

"**Minimal**" **problem** should be applied to visible or anticipated surface contamination of less than $10\,\text{ft}^2$.

"**Moderate**" **problem** should be applied to areas larger than $10\,\text{ft}^2$ of surface contamination but less than $100\,\text{ft}^2$; or to very heavy and continuous growth covering less than $10\,\text{ft}^2$.

"**Major**" **problem** should be applied to visibly mold-contaminated surfaces that are larger in area than $100\,\text{ft}^2$, or to very heavy and continuous growth covering less than $100\,\text{ft}^2$.

For "major" problems, the possibility of extensive hidden contamination should be explored, at least qualitatively, during the investigation phase, to inform remediation planning efforts.

Degree of Contamination

In addition to the surface area of mold growth, the density of mold growth should be considered in relation to the contaminated material and the potential for hidden growth. When hidden growth is possible or porous materials are contaminated, a higher problem category should be considered.

"**Minimal**" **problem** should be applied to situations where visible growth is mainly limited to scattered small colonies, and evidence suggests that these are only present on easily accessible nonporous surfaces.

"**Moderate**" **problem** should be applied to areas with light and spotty visible growth on porous or semi-porous materials, or when about half the area is covered by mold colonies on nonporous materials, or when other evidence (e.g., a short-term moisture problem or some mold odors) suggests that additional hidden contamination may be present.

"**Major**" **problem** should be applied to a heavy distribution (covering half the area) of many large colonies on any type of material, or when evidence suggests that hidden contamination is present and may be well established (e.g., knowledge of a long-term moisture problem and strong persistent mold odors).

Potential for Releasing Contaminants

Disturbance of contaminated materials by mechanical forces or invasive measures during remediation can release large quantities of mold particles to the air. The potential for disturbance of growth sites, which depends on the accessibility and nature of the material, should be considered.

"**Minimal**" **problem** should be applied to contamination on surfaces of items that can be cleaned easily in place, or to smaller removable items that can be bagged (enclosed in plastic) prior to removal.

"**Moderate**" **problem** should be applied to areas requiring removal of larger items as individual components with negligible disturbance, or to contamination that requires cleaning with average force or handling (e.g., scrubbing).

"Major" problem refers to situations where aggressive or destructive physical force (such as mechanized methods) will be needed for removal or cleaning of contaminated items, resurfacing, or gaining access to contamination. This category may apply to situations such as removing wallboard, partitions, and carpeting, or sanding wood or concrete.

Other situation-specific issues may indicate a need for additional health and safety precautions. School officials must use their best judgment to anticipate if, and how, any additional factors may impact risks, affect perceptions of important stakeholders, or influence costs. The following are examples of such factors that may need to be considered.

- The skill and experience of the individuals who will perform the work
- The presence of highly susceptible occupants or remediation workers
- The duration and scope of the remediation project
- The extent of public concern—it is prudent to treat the problem as a "major" mold problem if distrust and anxiety has developed
- Liability or other legal concerns
- The ability to control exposures through administrative or engineering controls

After investigating and discussing the amount of mold growth, the degree of contamination, the potential for releasing contaminants, and the other factors, school officials should determine which of the three mold problem categories ("minimal," "moderate," or "major") best fits their understanding of the problem. If information is lacking to apply a category to one of the three criteria, it is prudent to assume the problem is "major" until information is available to change that designation. If different categories are assigned to each criterion, it is prudent to apply the most cautious approach and pick the highest category. For example, if the amount of mold growth and degree of contamination are "minimal" but the potential for releasing contaminants is "moderate," then MDH recommends designating the problem as "moderate."

School officials must develop a remediation plan tailored to the specifics of the problem. Highly susceptible individuals include people with severe asthma, severe allergies, and compromised immune systems. Immune-compromised individuals include uncontrolled diabetics, people receiving immune suppressive drugs for organ transplants or cancer, leukemia patients, and people with immune deficiency diseases, such as advanced AIDS. A physician should be consulted about these issues.

▶ ADMINISTRATIVE CONTROLS

Administrative controls are decisions that can be made by school administrators to protect occupants from actual or perceived exposure to contaminants released during remediation activities. Controls may include measures such as removing or relocating occupants, and scheduling work during vacations, evening, or weekend hours. Practical and logistical considerations may also make it necessary to temporarily prohibit occupants from entering the work zone and possibly adjacent areas depending on the nature and duration of the anticipated remediation project. If a medical or other health professional recommends the removal of certain individuals for health reasons, school officials should attempt to accommodate such advice whenever possible. School officials should make sure that the area occupants are being relocated to is more healthful than the area they were moved from.

For both health and practical reasons, administrative controls should be considered for any mold remediation project. School officials should consider the following questions to determine whether administrative controls are needed.

- How concerned are occupants about being near the remediation site?
- Is it likely that a lot of mold contaminants or construction dust will enter occupied areas?
- Are any occupants medically known to be susceptible to molds (medically recognized sensitivities of greatest concern are asthma, mold allergy, compromised immune system, and hypersensitivity pneumonitis)?
- Are there any credible reports by occupants of adverse health effects attributed to the mold problem?

When evaluating the need for administrative controls during a remediation project, school officials should initially apply the most protective approach. Such a precautionary approach is usually prudent because the primary objective of any remediation project is to protect occupants' and remediation workers' health and there may be uncertainties, such as:

• The extent of the contamination problem has not been clearly identified.
• There is less than ideal confidence in the engineering controls to be used during remediation work.

Following careful consideration of the control measures and other remediation practices that should minimize occupants' exposures, school officials may choose to relax on the administrative precautions. For example, it is prudent to relocate susceptible occupants in areas adjacent to the mold remediation work area, until it is verified that the work area is under appropriate containment (such as following measurements and visual observation of negative pressure relationships between the work area and adjacent occupied areas).

It is important to clarify that the most protective approach is not necessarily the evacuation of an entire school building. Most of the worst mold problems can be remediated while utilizing engineering controls of the affected areas and possible evacuation of occupants or susceptible individuals in directly adjacent areas, while maintaining normal operations in the rest of the building. However, school officials may choose to close a building in extreme cases. This might apply when the entire building is contaminated (or significant airborne contaminants cannot be controlled) and a medical professional has linked significant health effects to the mold problem. School officials may consider closing a school for political or liability reasons, even though the availability of engineering controls and health-based scientific analysis does not warrant such an action.

Whenever occupants have been moved, school officials should anticipate questions about reoccupancy and safety after job completion. Post-remediation evaluation findings are necessary for making reoccupancy decisions. Communicating these findings is essential to provide peace of mind to the occupants. Implicit is the need to determine (in the planning phase) specific clearance indicators or criteria that will be used to evaluate the effectiveness of the remediation. It may be useful to include stakeholders in high profile or "major" contamination problems, since this can help anticipate questions and concerns that may need to be addressed and to manage their expectations. After reoccupation, occupants should be informed about the process for reporting any future concerns to the appropriate school official(s) (e.g., the indoor air quality coordinator).

It is understandable that school officials will be reluctant to involve certain stakeholders (e.g., certain affected parents or staff) because this may make discussions contentious and cumbersome. However, the views of these people will likely have to be addressed at some point. It is better to involve these people early on to minimize distrust and disappointment, and minimize the likelihood of performing additional work to please certain groups.

▶ REMEDIATION PRACTICES CONSIDERATIONS

Once the remediation scope and the communication strategy have been determined, and the administrative controls considered, school officials should then review the various possibilities and considerations relevant to the cleaning and removal of mold contamination. School officials may want to use the step-by-step example mold remediation procedures at the end of this chapter (see Boxes 7.5, 7.6, and 7.7), but should supplement these generic procedures with enough site-specific details to provide specific guidance to those performing the work (see Box 7.3).

Box 7.3 Preparing a Remediation Plan

Remediation project plans should cover the following topics at a minimum:

1. Remediation of excess moisture
2. Identification of possible hazardous materials (e.g., lead and asbestos) in abatement areas
3. Mold abatement practices and procedures
4. Health and safety precautions
5. Determination of project completion
6. Repair and reconstruction

Note: The data in this chapter provides limited guidance on 1, 2, and 6.

Detailed written project specifications should be developed, especially when the problem is "major," outside contractors are hired, or circumstances are high-profile and emotionally charged. Specifications should clearly define the responsibilities of all parties involved in the work. The requirements for removal, salvage, cleaning, abatement of other hazards, and any modifications or repairs should be clearly described.

A written procedure should be given and explained to the workers, and prohibitions should be emphasized (e.g., work hours, cleaning methods, chemical application, containment). Expectations for successful completion should be included in the project specifications. If measurable results are needed, then scientifically defensible criteria for successful completion need to be included. School staff planning remediation of mold contamination problems should carefully review the following subsections for applicability to their circumstances.

Deciding Whether to Use Outside Help

Once the nature and extent of the problem is understood, school officials have to decide whether school staff can adequately perform the remediation work or if outside assistance is needed. School staff can usually remediate "minimal" problems, and, if experienced, school staff may also be capable of remediating "moderate" problems. However, "major" problems should typically be remediated by experienced professionals who have the appropriate expertise and equipment, to protect workers and occupants and contain remediation areas. In addition, when district staff will remediate, it may be prudent to have a qualified professional review the mold problem and the work plan. There are two additional issues to consider before deciding to have school staff remediate the problem.

1. It is prudent for individuals who are immune-compromised, severely asthmatic, or suffering from hypersensitivity pneumonitis to avoid mold remediation work.
2. The MN Occupational Safety and Health Agency's (MNOSHA) Right to Know Act (Rule 52.06) must be followed as a part of training provided to maintenance staff regarding the potential health hazards associated with mold and any chemicals used for remediation. Without such training, school staff should not undertake mold remediation activities.

Personal Protection

Physical disturbance of moldy materials can produce extremely high airborne levels of mold particles and contaminated dust. When handling moldy materials or working in the remediation area, people should be protected from being exposed to contaminants. School officials should determine what personal protective equipment to require for in-house staff, depending on extent of the contamination and OSHA requirements. While there are no legal respiratory protection requirements specific to handling indoor mold, there are OSHA requirements that an employer must follow if they mandate employees use personal protective equipment.

For "minimal" to "moderate" mold problems that are manageable by routine custodial or maintenance activities, MDH recommends workers be provided with an N95 respirator, dust-proof goggles, and skin protection at a minimum. Remediation workers should be instructed to always wash their hands after working with moldy materials, in case any mold is unintentionally transferred to their hands. For some "moderate" and all "major" problems, it is prudent to use a high-efficiency particle air (HEPA) filtered full-face respirator. It is also important to select appropriate gloves. When applying a disinfectant or a strong cleaning agent is used, use gloves made from natural rubber, neoprene, nitrile, polyurethane, or PVC; when handling sharp materials, use leather gloves.

If other hazards (such as asbestos or lead) will be disturbed, properly licensed professionals must perform the work and follow the appropriate regulatory requirements. Cleanup of certain fungi, such as *Histoplasma capsulatum* or *Cryptococcus neoformans* that may be growing in bat and bird droppings require specific cleanup practices.

States with no state OSHA program must follow the federal OSHA Hazard Communication Standard (29 CFR 1910-1200). See CFR 29 1910.134 and 1926.95-107 for details. In situations where the employer does not mandate use, but the employee volunteers to use an N95 dust mask, employers must still follow some of these requirements. For information on statutorily mandated requirements, contact the Minnesota OSHA Workplace Safety Consultation unit at 800-657-3776 or 651-284-5060.

Particulate respirator names refer to the ability to remove oil mists (N: not oil resistant; R: oil resistant; P: oil mist proof), and the filtering efficiency of particulates > 0.3 μm (95, 99, or 100%). If oil mists are present, "P"-rated respirators should be used. These respirators will not remove some odors since the gases released from mold may cause the odors. If odors are a significant concern during remediation, respirators with organic vapor cartridges may also need to be used; see the DHHS/NIOSH Publication No. 2005-109.

Contaminant Control

Remediation plans should include project- and site-specific instructions on how workers will minimize and contain the release and spread of mold particles to any occupied or non-contaminated areas. Contaminants may be spread by foot traffic, contaminated materials or equipment, or through air movement. More stringent contaminant control methods are necessary when large amounts of mold growth are being remediated or destructive techniques are used. Workers should handle contaminated materials in a manner that minimizes disturbance of fungal particles, especially if their removal might uncover further contamination that has not yet been identified. To prevent or minimize the dispersion of particles beyond the work area, containment (i.e., engineering controls) and special cleaning practices are often necessary. This may include critical barriers (e.g., polyethylene sheeting), depressurization techniques (e.g., negative air pressure machines), dust suppression methods (e.g., damp wiping and HEPA vacuum cleaning), and decontamination procedures (e.g., showers, dirty rooms, clean rooms). Step-by-step recommendations are given at the end of this chapter (see Boxes 7.5, 7.6, and 7.7). For further guidance on contaminant control, school officials should seek advice from experienced mold remediation professionals.

In "moderate" and "major" problems, workers should "preclean" areas to minimize dispersion of contaminants while assembling engineering controls and applying aggressive force to contaminated materials. Precleaning involves careful cleaning of easily accessible surface contamination with methods that can lift and capture fine dusts. Where possible affected materials should be enclosed in plastic and removed from the building using the shortest direct route leading to the outside of the building.

Containment should be designed according to the severity of the problem (see the previous section, Determining Remediation Scope). During remediation of "moderate" contamination, it may be adequate to enclose an area in plastic sheeting (a localized mini-enclosure or tent) and then bag and dispose of contaminated items. In the case of "major" contamination problems, physical isolation (total enclosure including at least one decontamination chamber) and depressurization may be necessary to separate work areas from noncontaminated spaces and the heating, ventilation, and air conditioning.

Depressurization of the work area is often used to ensure contaminants are confined within the work area. A pressure differential of > 5 Pa (0.02 inches water column), which is used for asbestos abatement, is usually adequate to contain airborne mold contaminants. In addition, remediation workers and project managers should routinely check to see that the separation barrier, typically plastic sheeting, billows inwards into the isolated area. The nozzle of one or more HEPA vacuums drawing from within a small containment area may provide adequate depressurization. When larger areas are isolated and a greater volume of air must be filtered, HEPA-filtered negative air machines are necessary.

Cleaning and Removal of Contaminated Materials

The primary goal of effective mold cleanup is to capture and remove contaminants from the building. Sometimes the material that the mold has grown on must also be removed. As a general rule, it is best to clean and remove as much contamination as possible before the materials have completely dried. After any mold cleanup, it is necessary to periodically inspect areas of previous mold growth to ensure the problem has been corrected. The following guidelines should be considered for determining which materials can be cleaned and salvaged versus which should be discarded.

Discard Porous Materials with Evidence of Mold Growth

Evidence of mold growth consists of visible mold or mold odors emanating from the material. Examples include paper fiber gypsum board, ceiling tiles, insulation, wall coverings, carpet, leather, and unprotected composite or engineered wood products. Such items are not easily salvaged and it is not often cost-effective to remove the mold. Instead, it is usually best to bag or wrap the materials in plastic and discard. There may be instances where small areas of superficial mold growth on a porous material can be cleaned. For example, mold may be growing on a superficial organic nutrient, such as a spilt soda. Such growth can be cleaned with a HEPA vacuum, followed by water extraction, rapid drying, and finally HEPA vacuuming again. However, when in doubt it is best to replace the material. Interior lined ductwork is also porous. If ductwork lining is colonized by mold, it should be removed and the duct cleaned down to bare metal or simply replaced.

Porous Materials without Evidence of Mold Growth

What should you do with porous materials without evidence of mold growth? Although not visibly moldy themselves, many rough-surfaced porous materials may also become reservoirs of settled fungal particles if they have been near heavy growth or high air concentrations. Examples include upholstery, carpet, modular furniture, books, and files. These materials should be cleaned by careful HEPA vacuuming, laundering, or other methods that lift and capture fine dusts from the material's surface.

Semi-Porous Materials with Little Surface Growth or That Are Structurally Sound

How do you handle semi-porous materials that have little surface growth or are structurally sound? Examples include solid wood furniture or structural components, protected composite and engineered wood products, concrete, cement, brick, and some resilient floor coverings. Such items may be salvaged if they are structurally sound and can be kept dry in the future. Cleaning and remediation steps include one or more of the following: HEPA vacuuming, damp cleaning with soap and water, HEPA vacuuming again, resurfacing (in the case of wood), disinfecting, drying rapidly, and sealing or refinishing. If the above cleaning methods are ineffective a HEPA vacuum sander (simultaneous vacuuming and sanding) can be used, but do not attempt to sand without simultaneously using a HEPA vacuum. When concrete surfaces cannot be cleaned by HEPA vacuuming or detergent and water, trisodium phosphate (TSP) can be used to clean concrete. TSP is a strong irritant and must be used with caution.

Semi-Porous Materials with Extensive Fungal Penetration

Examples include solid wood furniture or structural components, protected composite and engineered wood products, studs, paneling, and some resilient floor coverings. When extensively colonized, such items are not easily or cost-effectively cleaned and decontaminated. They should be removed, bagged, and discarded. If contaminated materials cannot be removed and the structural integrity of the material has not been jeopardized, it may be acceptable to clean and disinfect rigorously and take exceptional measures to prevent the return of moisture. In addition, it is critical to remain vigilant for signs of mold problems in the future and to respond rapidly if they return.

Nonporous Materials

Examples include metal, ceramic tile, porcelain, glass, hard plastics, highly finished solid wood items and other hard, smooth, nonpermeable surfaces. Cleaning steps include HEPA vacuuming, damp wiping with a detergent solution, and rapid drying.

Moisture Control

The importance of addressing moisture in any effort to solve a mold problem cannot be overstated. The presence of excess moisture is the primary underlying cause of indoor mold growth. Identifying and correcting sources of excess moisture is vital to resolving mold problems and preventing their reoccurrence. Judging the completion of a remediation job should include evaluation of steps taken to correct moisture problems and prevent their return. If a catastrophic water problem occurs, it is critical to quickly correct the water source and to initiate restorative drying practices before mold growth occurs or spreads. Building materials should be dried rapidly, ideally in less than 48 hours, preferably less than 24 hours, to a moisture content that does not support mold growth. Check that there is no visible mold growth before drying areas, because restorative moisture drying methods can pressurize materials (such as wall cavities) allowing mold particles (if present) to spread to uncontaminated areas.

Tools used for drying include extraction, evaporation, dehumidification, and temperature control. Nonsalvageable items that have been wetted or begin to show evidence of mold growth should be discarded promptly. As a precaution, fans and other devices that create airflow should no longer be used once visible mold appears, and drying should then proceed under more controlled conditions to avoid dispersing mold particles. Moisture source(s) that have led to mold growth must be identified and understood to plan an effective mold remediation project. Finding the locations of excess moisture, identifying the mechanisms for its accumulation or infiltration, and tracing pathways of its migration can also aid in finding and assessing the likelihood of encountering further growth.

Staff responsible for correcting and preventing mold problems should recognize that moisture in any of its phases (ice, water, and vapor) must be adequately controlled. For example, moisture may be present at a material's surface as high relative air humidity. Moisture can also be absorbed into porous and semi-porous materials and may migrate under surface coverings, around furniture, and between components at joints. Understanding the moisture sources and dynamics involved in a particular situation may require professional assistance.

During the mold remediation project itself, it is necessary to control the use or production of water. For example, cleaning techniques should use water-based solutions sparingly and must include rapid drying following the cleaning steps. Power washing should be used as a last resort, and only on nonporous materials or concrete (cementitious materials) if the material can be dried quickly. Power washing should not be performed if vulnerable material, such as wallboard and Sheetrock, might get wet. In addition, significant amounts of moisture can be introduced into air from open flame heaters. These heaters should be avoided in remediation areas.

Mold will likely grow again where moisture issues are not addressed adequately. School officials planning and carrying out mold remediation should consider how to ensure and document that moisture problems have been resolved. See the section Remediation Evaluation for details.

Cleaning of Remediation Equipment

Equipment used during remediation, such as respirators and protective clothing, may need careful cleaning depending on how much mold was released during cleaning. In the case of a "minimal" problem, tools and personal protective equipment can usually be adequately cleaned by damp wiping or washing with soap and water. With "moderate" and "major" problems, a protocol for decontaminating workers and equipment should be developed. In the case of "major" problems, containment should be constructed to include a separate decontamination chamber (with plenty of room to work comfortably).

All equipment should be HEPA vacuumed, damp wiped, and bagged before removal from the work area. This includes cleaning tools, negative air machines, waste containing bags, outer clothing, respirators, gloves, and goggles. Workers should wear at least an N95 respirator when cleaning or replacing HEPA-filtered equipment components. At the end of the removal effort, all materials used for containment should be bagged and the area decontaminated as part of the final job site cleaning. If hazardous materials such as lead or asbestos are also handled as part of the removal work, applicable regulatory work practices and procedures must be followed.

Waste Disposal
Mold contaminated materials are not classified as hazardous waste, and they can be disposed of in a sanitary landfill. However, mold-contaminated waste that is not immediately disposed of should be stored securely (e.g., in a covered and posted waste container) and located away from high traffic areas, entrances, and fresh air intakes. Any hazardous materials removed must be kept separate from the nonhazardous waste, labeled appropriately, and disposed of according to applicable rules and regulations.

Heating, Ventilation, and Air-Conditioning Systems
The term *HVAC system* refers to the entire air distribution system from points where air enters the system to points of discharge. This may include return plenums, ceiling plenums, and the mechanical room. The heating, ventilation, and air-conditioning system(s) is relevant to mold remediation because it may be the source of mold growth or the route of disseminating mold particles from one area to another. Without regular inspection and proper maintenance of critical HVAC system components, dust, debris, and moisture may collect beyond the usual amounts expected. And since some HVAC system components may be inaccessible

for periodic inspection, they are particularly susceptible to mold growth. The remediation of school HVAC systems is generally similar to those for other building components or systems, although additional precautions and hiring a professional may be necessary.[1]

An evaluation of the HVAC system should be done for any severity of mold problem. In particular, the entire HVAC system should be assessed for its role in the moisture problem(s). However, such an evaluation can be skipped for suspected "minimal" problems (due to the expense of hiring a ventilation professional) when the HVAC system is not accessible and when there are no signs or indications of HVAC system involvement. In some cases, a reservoir of spores or location of growth may be the acoustical insulation sometimes used to line interior airstream surfaces of air-conveyance ducts. If such lining (or any other nonsmooth or porous airstream surface) is colonized by mold growth, it should be removed, discarded, and cleaned down to bare metal. Unit ventilators should not be overlooked as potential sites of mold contamination, since they are often poorly maintained and their operation hampered by misuse as storage areas.

The following HVAC system components should be inspected for growth, moisture, and relevant defects, and cleaned or replaced, as needed.

- Outdoor air intakes
- Filters
- Cooling coils, including evaporator fins
- Condensate pans, collectors, and drains
- Humidifiers
- Air stream surfaces (baffles, dampers, including internal acoustical lining, fiberglass duct board, etc.)
- Blowers, fan components, and housings (supply, return, and exhaust)
- Air distribution devices (registers, grilles, and diffusers)

---

[1] For example, for "moderate" problems in HVAC the New York City Department of Health and Mental Hygiene recommends constant negative pressure and particulate collection (e.g., vacuum collectors) in addition to the other considerations. For problems that cover more than $30\,ft^2$, NYC also recommends full-face HEPA respirators, airlocks, and decontamination rooms.

MDH advises against the routine use of disinfectants and pesticides to "sanitize" ducts in school HVAC systems (see the next section, Use of Disinfectants and Pesticides). If cleaning and removal is done properly, disinfectant or pesticide application should be unnecessary.

Furthermore, if organic materials (i.e., mold, dust, debris) are not removed first, it is unlikely that disinfectants or pesticides will achieve their intended level of killing anyway. The health effects of such chemical use in HVAC systems are poorly understood, and improper application may lead to greater problems or complaints than the mold. Currently, there are few such chemicals that can legally and practically be used in ductwork, and there is a lack of compelling evidence regarding their effectiveness in field applications.[2] If school officials choose to use chemicals in their HVAC system, MDH cautions that such agents should never be applied in an operating HVAC system, and the manufacturer(s) of components that will be treated should be consulted before treatment to ensure compatibility.

MDH does not currently support the use of sealants or encapsulants as the primary remediation strategy to address mold problems in HVAC systems. There are important concerns about the toxicity and odors associated with sealants and encapsulants, and there are uncertainties about their long-term success in containing mold particles. While such a strategy is not a substitute for removing mold contamination or porous materials subjected to excess moisture, sealing or encapsulating colonized materials may be preferable to taking no action. If moldy materials are "locked down" in place, MDH recommends keeping a record to ensure appropriate care is taken to minimize the potential of future disturbance.

Use of Disinfectants and Pesticides

While usually unnecessary, disinfectants or pesticides may be used to try to kill any remaining mold following cleaning or material removal. Many conventional cleansers have disinfectant

---

[2] See the U.S. Environmental Protection Agency's "Should You Have the Air Ducts in Your Home Cleaned?" for more information (*www.epa.gov/iaq/ pubs/airduct.html*). The recommendations in this document also apply to institutional buildings.

properties, such as household bleach, and are subject to limited regulations. Pesticides, on the other hand, are specifically regulated. Before using a disinfectant or pesticide, school officials should consider the following issues.

- Whether the situation truly necessitates its use, or are other methods sufficient
- Whether the agent has been effective at controlling the target microbe, in schools, and on the same materials, when applied at levels that have negligible health risks
- Whether they are applying a disinfectant or pesticide
- If it's a pesticide, whether the agent will be applied by an experienced and licensed applicator, according to laws and safety guidelines

It is very important to review the labels of any cleaning or chemical agent used in mold remediation, to determine whether it is regulated as a pesticide and to ensure proper application. The term "biocide" refers to a legally defined group of pesticides used for managing microbial pests (such as mold). In Minnesota, pesticide applicators must have an antimicrobial commercial license to apply pesticides used to kill mold in schools. It is a common misconception to think that because a pesticide is "EPA registered" it is safe to use in a variety of situations—pesticides are registered only for the specific use(s) described on the label. Contact the Minnesota Department of Agriculture for more information on pesticide regulations, applicator licenses, or other specific questions. If applicable, the school's integrated pest management plan should be followed.

MDH's primary concern is for occupants' and remediation workers' health and safety. Even when handled according to label instructions, any disinfectant or pesticide should be considered potentially harmful to people, and, rightly or not, such agents may be blamed for health problems and odors. In addition, even if an agent does kill the mold, the agent does not usually destroy the allergens, irritants, or toxins that are present in the dead mold. Hence, dead mold particles from contaminated materials can still adversely affect health. Consequently, it is of questionable value to kill mold before removing it, but killing the mold without also removing the contaminants is not acceptable.

There may be circumstances in schools where judicious use of a disinfectant or pesticide could be beneficial to kill residual living mold growth, but only after cleaning and removal efforts have removed visible mold growth. First, MDH recommends the use of disinfectants or pesticides if severely immune-compromised individuals[3] are expected to occupy or reoccupy a previously moldy area. These individuals are susceptible to infection with the living mold spores—it is therefore useful to kill any residual mold that may still be present following cleaning and removal of moldy materials. Second, a disinfectant may be helpful on stubborn mold problems growing on nonporous or semi-porous surfaces that cannot be replaced (e.g., concrete blocks).

Standard household bleach (i.e., hypochlorite) is often used to clean and disinfect materials. Some hypochlorite solutions are regulated as pesticides. Bleach should not be used on materials that will corrode, for example stainless steel. For initial use, dilute the stock solution (if 5% hypochlorite) 20 parts water to 1 part bleach to yield a 0.25% hypochlorite solution. Since many agents, including bleach, are rendered ineffective after reacting with microbial contamination or other organic soiling, they should be applied only to previously cleaned surfaces using clean applicators (buckets, mops, sponges, etc.) or dedicated equipment. Apply the solution with a damp cloth and leave it on for a period of time according to the manufacturer's direction (some disinfectants should be left on for 10 minutes or less, while bleach is usually left on for 30 minutes). After disinfecting, the remaining chemical residue should be damp wiped from the treated surface with clean water, and the material should be dried quickly.

Bleach should never be mixed with ammonia-containing products nor should it be applied to a hot surface—both will produce toxic chlorine gases. Bleach should only be mixed with other chemicals if this is permitted on the label. Since bleach and most disinfectants and pesticides are volatile chemicals, they should only be applied when adequate ventilation and

---

[3] Immune-compromised individuals include uncontrolled diabetics, people receiving immune suppressive drugs for organ transplants or cancer, leukemia patients, and people with immune deficiency diseases, such as advanced AIDS. Individuals should contact their physician for more information.

appropriate respiratory protection are used. When bleach is handled, the respiratory protection equipment used must be effective against inorganic vapors. In addition, protective gloves and eye protection should be used when handling bleach to avoid burns.

Treating moldy materials with disinfectants or pesticides can also complicate efforts to evaluate cleanup. Treated materials may still release large numbers of mold particles that will not be measured by culture-based (i.e., viable) sampling and analytical methods. Evaluation of a remediation job should not rely solely on viable testing methods if disinfectants or pesticides were used. If viable sampling is used, it should be done after cleaning, but before disinfectant or pesticide application.

Use of Gas-Phase Ozone

Ozone-generating devices have been promoted to Minnesota schools as a solution to mold and other indoor air quality problems. There are significant concerns about applying ozone, a well-known respiratory health hazard, to occupied indoor environments. The limitations of using a disinfectant or pesticide described above also apply to ozone. In addition, as a general rule gas or vapor phase agents cannot effectively and safely remediate microbial building contamination—effective and safe application typically requires direct contact with the contamination. Because ozone may eliminate or mask odors, it can create the perception that the mold problem has been resolved. Yet research, including recent controlled laboratory studies, has confirmed that gas-phase ozone, applied at levels below health standards, is not effective at inactivating microbial contamination (including a variety of fungal organisms) on building materials or in air. Consequently, MDH strongly opposes the intentional use of ozone to address school mold problems.

▶ REMEDIATION EVALUATION

After the mold remediation work is completed, an evaluation of its effectiveness should be performed. At a minimum, school officials should verify and document that moisture problems were corrected and the contaminated area is ready for occupancy. Evaluation findings can also help restore the confidence of all stakeholders. Outside contractors should be given a clear description of job requirements, including how successful completion will be determined. School officials and their contractors may also desire an independent third party's evaluation as a way to provide some protection against potential future liability. School officials must decide who will do the evaluation, how the results will be used to demonstrate that remediation goals have been met, and what next steps will be taken if clearance criteria are not satisfied.

Remediation Goals

Clear and achievable goals should be set during remediation planning. All parties involved in the project should understand and agree on the goals. It may be worthwhile for other stakeholders and affected parties to participate in setting goals, since they may better appreciate the costs and difficulties associated with expectations that are unrealistic and impractically strict. The ideal remediation goal is to restore the building to conditions in which occupants are free from health complaints or discomfort. It is, however, extremely difficult to achieve and maintain such a level of satisfaction given the many agents and conditions that can contribute to real and perceived indoor air quality problems and complaints. Some may demand that the goal should be a building free of all molds, but this is not possible or practical since, at a minimum, spores will always be detectable in settled dust and in the air.

A reasonable remediation goal is to restore the building to normal conditions, reflected by: a lack of visible mold growth; a lack of mold odors; and appropriate control of moisture. Another goal in some cases may be to confirm, through testing, that the types and amounts of mold particles in the air or settled dust are similar in type and amount to what is present in unaffected and outdoor areas. Finally, remediation goals should also fit into a holistic approach to improving and maintaining indoor air quality through preventive maintenance, rather than simply reacting to problems.

Evaluation Criteria and Methods

Once goals for the remediation have been determined, evaluation criteria and methods can be selected. The methods and extent of evaluation should depend on several factors, especially the extent of the contamination problem and the community's concerns.

For "minimal" contamination cleaned by routine housekeeping, a sensory inspection alone should be adequate to judge project completion. When "moderate" or "major" contamination is present or health concerns have elevated the importance of the issue, a more thorough evaluation and communication of findings are advised.

Setting evaluation criteria involves determining ahead of time how much contamination may remain after remediation is complete—in other words, deciding what indicators or measurable results will be considered evidence of an acceptable outcome or job "clearance" (see Box 7.4). These criteria need to be set before remediation work begins. Setting clearance levels too low will impractically increase costs without additional practical benefit. Instead, the criteria should be selected to show, in combination with other evidence and information about the remediation activities, that the remediated area was acceptably clean and dry at the time when the job was finished and that conditions that allowed mold growth were corrected. When using numerical criteria for clearance, it may be necessary to set material-specific and test method-specific criteria for interpret-

---

Box 7.4 Evaluation of Remediation Methods

The project manager should confirm with remediation workers that the previously determined remediation plan was followed. This should be verified during and after the remediation work by periodic inspections and closeout documents that should be included in the final report. School officials may wish to perform this task or delegate the responsibility to a contracted professional. The following are examples of some common problems that the project manager may identify:

- Incorrect mixing of chemicals
- Inadequate precleaning before moving large items from or erecting walls in remediation area
- Overloaded HEPA filters in vacuum cleaners and negative air machines, which no longer draw enough air, have insufficient negative pressure established, or a loss of negative pressure (apparent on manometer readings)
- Complacency regarding use of respirators
- Not bagging materials and cleaning off surfaces before leaving containment area

If significant inadequacies are revealed, proper remediation should be resumed before remediation activities continue.

ing testing results. This must be determined before the remediation work is begun and should ideally be understood and
accepted by all key stakeholders.

Sensory Approach

The sensory approach should be used to evaluate all mold remediation efforts, from the most routine "minimal" problems to
"major" problems. The sensory approach involves using senses
of sight and smell to determine the presence or absence of visual
and olfactory signs of conditions that support mold growth.
Combined with evidence that effective methods for removing
mold contaminants were used and moisture problems were
addressed, the sensory approach offers a practical and commonsense option for evaluating whether remediation goals have been
met. Sensory criteria should include, at a minimum, that there
is no visible mold growth, negligible dust, no moldy odors, and
no apparent dampness.

One very important indicator of mold removal effectiveness
is the overall cleanliness of the work site after job completion.
The presence of any remaining visible mold colonies indicates
that cleaning and restoration was not adequate. Moreover, the
presence of dirt, moisture, debris, and dust should not be toler-
ated in remediated areas after project completion. Methods to
document a sensory evaluation include written testimonials,
photographs, white glove/black glove inspection for dust, and
confirmation by an independent third party. A white glove/black
glove inspection involves allowing suspended matter time to
settle, then wiping a finger over all or representative (previously
determined) surfaces to demonstrate general cleanliness. In
addition to the areas of contamination, areas that will underlie
repaired or reconstructed structures should be assessed (cleared)
before reconstruction, to verify contamination and moisture
problems have been properly addressed.

Moisture Testing

In addition to the sensory approach, one common method for
evaluating moisture control is to measure moisture levels of
affected materials and surfaces and/or the indoor air's relative humidity. Criteria for acceptable moisture levels in air
or specific to certain materials should be established early in

the project. These criteria can be used to verify that adequate drying has occurred before the replacement of damaged materials, refinishing, installation of surface coverings, or other reconstruction efforts. In addition, they can be used to verify whether a moisture control method has been truly effective at controlling moisture. Equipment such as moisture sensors or detectors, thermohygrometers, and moisture meters may be used to evaluate drying progress and confirm moisture control. School officials should be familiar with the terms defined in Table 7.1. Moreover, they should expect contracted professionals who test for to be very knowledgeable about these concepts and their application to investigations and restoration.

When verifying acceptable moisture levels it is preferable to compare moisture measurements to published acceptable moisture content (MC) values for a particular material relative to a benchmark water activity (a_w). As a rule of thumb, an a_w of less than 0.65 is ideal because this is typically the minimum amount of available water necessary for microbial growth. Microbial growth is limited at a_w up to 0.75. Note that a specific a_w value

TABLE 7.1 Terms Used to Describe Water in Air and Materials

MEDIUM	TERM	DEFINITION
Air	Absolute humidity	The ratio of the mass of water vapor to the mass of dry air
Air	Relative humidity (RH)	The ratio of the amount of moisture held in air to the maximum amount that the air can hold for a specific temperature and pressure
Materials	Moisture content (MC)	The mass of moisture held in a material, measured as the mass water as a percentage of the dry mass of a material
Materials	Water activity (a_w)	The ratio of the amount of water in a material at a particular moisture content to the maximum amount of water air can hold at the same temperature and pressure
Materials	Equilibrium RH (ERH)	The a_w expressed as a percentage

Note: From *ACGIH Bioaerosols: Assessment and Control.* See Appendix B of this report for further information on moisture assessment and control.

corresponds to very different measured MCs, depending on the material. For example, an a_w of 0.75 corresponds to about 20% MC for pine wood and 5% MC for concrete.[4] If published moisture content and corresponding a_w are not available for a specific material, then background measurements from similar dry materials may be used for comparison to confirm a return to normal moisture levels. The use of relative humidity may be useful if measurements are taken directly next to the suspected material. However, RH measurements may fail to identify cold spots or local water intrusion that has caused a high localized a_w enabling mold growth.

Mold Testing

When done in a scientific manner, mold testing can provide useful information that complements other post-remediation evaluation approaches, described above. School officials should try to understand the limitations, uncertainties, and nuances of mold testing to determine: whether it is a good investment of limited resources; whether it can answer the desired questions; and if it can be done in an objective, useful, and defensible manner. Mold testing is not something school officials should feel compelled to do in any and every mold remediation project. For example, mold testing is usually unnecessary for "minimal" problems. Moreover, when testing cannot answer the desired questions or a scientific protocol cannot be afforded, it may suffice to rely on the other evaluation methods.

MDH strongly recommends that mold testing be performed only if it can be done adequately to answer a question with acceptable certainty. In general terms, the questions for post-remediation evaluation testing should be along the lines of "does the level of biological agents detected on this particular material indicate acceptable remediation?" or "does this environment contain more organisms than would normally be expected?" Based on the available science and lack of consensus standards, currently it is MDH's view that questions such as "is there a safe level of mold?" or "is the kind of mold present more harmful than others?" cannot be answered with acceptable confidence or certainty.

---

[4] From Janet Macher (Ed.), 1999. *Bioaerosols: Assessment and Control.* ACGIH.

Currently there is no widely accepted protocol for mold testing although a number of ways to estimate mold levels exist. Considerable expertise is needed, in most cases by an outside consultant, if sampling and interpreting results are to be done defensibly. Depending on the situation, type of contaminated material, and sample collection and analysis method, there are limits to what can be detected and quantified. For example, viable samples detect a portion of the live organisms present but provide no objective information on the amount or type of non-viable microorganisms that may also be present. This may be an important distinction if the predominant cause of complaints was nonliving mold particles, or if remediation killed the mold but failed to adequately remove it. Further information about mold testing is available in MDH's "Best Practices for Mold Investigations in Minnesota Public Schools"[5] and the MDH fact sheet "Testing for Mold."[6]

Because mold contamination is not always visible, mold testing can serve an important and necessary role in evaluating remediation. Sampling can also be used to assess the possible spread of contaminants from a containment zone to adjacent areas during or after remediation. In cases involving extensive contamination or high-profile situations, sampling has been used for post-remediation clearance when outside remediation contractors or consultants insist on testing. If sampling was done to investigate the problem, it may be prudent to test after the remediation because stakeholders may expect to see a significant decline in mold levels (however, changes in the molds' metabolic state or environmental conditions can also affect measured mold levels).

The costs of mold testing can be considerable since large numbers of many different types of samples are usually needed. Enough samples should be collected to allow statistical characterization of the likely variability in sample results. Mold colonization and deposition onto surfaces may be discontinuous, and releases of mold into the air may be intermittent. Sample results can vary dramatically depending on time and location, and many samples are needed to account for this variability.

---

[5] *www.health.state.mn.us/divs/eh/indoorair/schools/remediation.pdf*
[6] *www.health.state.mn.us/divs/eh/indoorair/mold/moldtest.html*

Depending on the nature of your question, different sampling methods are needed. If the question presented is whether a material has returned to a normal mold level, then preference should be given to a technique that determines if surface or dust levels pass or fail a specific clearance criterion. If a broader question is presented, such as whether mold levels in the overall environment (air and surfaces) has returned to normal, then typically three types of methods should be used:

1. Spore trap screening of air for total mold (viable and nonviable)
2. Sampling air for viable mold levels
3. Sampling surfaces for viable mold

If only two of the three methods are affordable or available, MDH recommends aggressive spore trap and surface sampling—these are likely to yield more information about the total fungal load or the location of contamination than viable air samples alone.

After samples are collected, they are sent to a lab for analysis. The analytical method should be sensitive to low levels of a broad spectrum of environmental fungi, provide the level of taxonomic identification needed for the intended use of the data, and should not be so expensive to deter collecting an adequate number of samples. When the types and quantities of mold are essential for interpretation (for example, comparing amounts and types of mold in air samples to background or outdoor samples), MDH recommends using a laboratory accredited by the American Industrial Hygiene Association's Environmental Microbiology Laboratory Accreditation Program (EMLAP).

Interpretation of mold testing results is somewhat subjective because there are no widely agreed on standards or criteria for acceptable levels of mold. This is primarily because there is no scientific consensus regarding an environmental concentration of mold that correlates with clear and consistent health risks. Therefore, it is overly simplistic and inappropriate to rely solely on a comparison of test results to any published numerical standard or value to determine safety or health risks. Some numerical guidelines have been proposed and published for use as "screening guidelines" or "performance criteria" to determine unusual versus expected amounts of mold. However, these values are specific to sampling and environmental factors such as: equipment make and model; culture media used; flow rate;

sampling duration; operator handling and performance; building operations; and geographic location and season. Comparison of results to such values should always consider differences in the sampling method, techniques, and environmental factors.

A superior alternative to interpreting air samples results strictly according to a published absolute numerical value is to compare air fungal estimates from problem areas to those from similar suitable background areas (i.e., unaffected indoor area air levels and outdoor air levels). This comparative approach tries to determine if the concentrations and diversity of molds present in the remediated area are similar to the outdoor and unaffected indoor area air levels. The following general principles should be used when interpreting comparison sampling results.

Air Samples

Comparison is only valid between samples taken at similar times on the same day and using the same sampling method (e.g., flow rate, duration, culture medium). Some variation in the total fungal levels and the presence or absence of a few types from one sample to the next is expected. Where relevant, indoor areas should be sampled and compared when building operations are similar, such as ventilation, open windows, cleaning and occupant activity level prior to and during sampling, and weather conditions. The following suggests acceptable mold levels:

- Total concentrations of mold (number of colony forming units and/or total spores detected per unit volume of air) in indoor samples should be similar to, or lower than outdoor and unaffected indoor area samples.
- Indoor samples consistently contain types of mold present in the outdoor and unaffected indoor area samples.
- Indoor samples are not dominated by the types of mold (as a percentage of the total amount) unless the same types also dominate the comparable outdoor and unaffected indoor area samples.[7]

---

[7] If sample results are to be interpreted by comparison to background levels, MDH suggests species level identification because this permits greater confidence in interpretation. However, due to cost considerations, it may be adequate to distinguish and compare fungi at the genus level. Identification of many species requires viable sample techniques. The results should be considered "presumptive."

Surface and Dust Samples

These should generally show similar levels and types of mold fragments expected on similar materials reported in the literature or measured on the same materials in unaffected indoor areas.

- Total concentration of mold (number of colony forming units per unit area or gram of dust) in indoor samples should be similar to, or lower than, samples from the same kind of material in an unaffected area and/or what is reported for similar materials in the literature.
- Samples should show a mixture of mold types, not be dominated by a single type of mold unless unaffected area samples are also dominated by the same mold.
- Microscopic examination of samples should indicate an absence of colony structures (spore producing structures and mycelial fragments) that indicate surface growth.

▶ EXAMPLES OF STEP-BY-STEP REMEDIATION PROCEDURES

The following are examples of step-by-step procedures on contaminant removal, engineering controls, and personal protection for three categories of mold problems ("minimal," "moderate," and "major"). As described in the section "Determining Remediation Scope," problem categorization is based on the amount of visible fungal growth observable on surfaces, the degree of contamination, and the potential for releasing contaminants. The procedures shown in Boxes 7.5, 7.6, and 7.7 are for instructional purposes. The practices and procedures for each specific mold remediation project need to be defined according to the many intricate variables that are too complex to fully address here.

The procedures described do not detail moisture control measures, which are critical to permanently address any mold problem. All the information in preceding sections should be read to fully understand all aspects of mold remediation. Professional evaluation or remediation services may be helpful or necessary—their work plan should be compared to MDH's and other reputable sources' recommendations.

Box 7.5 Practices and Procedures for "Minimal" Problem Remediation

Step 1. Select personal protective equipment. Workers should protect themselves with the following:
- Respiratory protection capable of filtering particles down to 1 micron (an NIOSH-approved N95 filtering face-piece respirator is recommended at a minimum)
- Eye protection (goggles that exclude fine particles)
- Gloves (impervious to any cleaning products used)

Step 2. Determine whether contaminated materials can be cleaned or whether they need to be discarded.
- Porous materials (including drywall board, ceiling tile, insulation, unprotected "manufactured" or "processed" wood products, upholstered furniture, carpet, and padding) that are contaminated with mold need to be removed from the building. This should include all materials and furnishings that have, or had: visible mold growth; strong mold odors; or remained wet for longer than 48 hours and are not easily cleanable.
- Hard-surfaced semi-porous materials such as tile, finished wood products, cement, and concrete can often be left in place and cleaned, if they are structurally sound, would be very difficult to replace, are only lightly contaminated on the surface, and can be successfully cleaned.
- Nonporous materials need to be thoroughly cleaned (includes metal, ceramic tile, porcelain, glass, hard plastics, highly finished solid wood items, and other hard smooth surfaces).

Step 3. Carefully clean mold contaminants by trapping or capturing as much of the visible mold growth as possible from accessible surfaces. Use component removal methods* whenever feasible; otherwise, select one or more of the following techniques.
- Vacuum all visible mold growth and materials surrounding the area of growth using a HEPA vacuum (a standard shop vacuum is not adequate)
- Carefully and systematically damp wipe surfaces with soapy water to remove and capture surface growth (work damp, not wet)
- Bag or contain porous contaminated materials, remove from work area

Step 4. Perform a final cleaning and drying of nonporous surfaces, including surfaces surrounding discarded porous materials.
- Damp wipe the cleaned materials with clean water to remove any remaining contamination or soiling residue.
- Manage run-off and leave surfaces as dry as possible after cleaning.

Step 5. Clean surrounding areas as needed.

Step 6. Perform post-remediation evaluation to determine the effectiveness of remediation work, and document the findings.

Step 7. Periodically inspect for moisture and visible mold growth.

* Component removal techniques involve enclosing or sealing the surfaces of whole assemblies or sections of building materials or furnishings in plastic or other impermeable materials before removal. For example, wrapping, removing, and disposing of entire components of cloth cubicles or entire sheets of wallboard in one piece.

Box 7.6 Practices and Procedures for "Moderate" Problem Remediation

Step 1. Select appropriate personal protective equipment. Workers should protect themselves with the following:

- Respiratory protection capable of filtering particles down to 1 micron (an NIOSH-approved N95 filtering face-piece respirator may be sufficient; a HEPA-filtered respirator (for example, P100) is strongly encouraged, if available, or if heavy disturbance is likely
- Eye protection (goggles that exclude fine particles) if half-face respirator is used
- Protective covering (disposable or washable outer clothing, long sleeved tops, long pants, booties, and head coverings)
- Gloves (impervious to any chemicals used, and, if applicable, protective against sharp objects)

Step 2. Determine if the material(s) supporting surface mold growth can be cleaned or should be removed and discarded.

- Porous materials (including drywall board, ceiling tile, insulation, unprotected "manufactured" or "processed" wood products, upholstered furniture, carpet, and padding) that are contaminated with mold need to be removed from the building. This should include all materials and furnishings that have, or had: visible mold growth; strong mold odors; or remained wet for longer than *48 hours* and are not easily cleanable.
- Hard-surfaced, semi-porous materials (e.g., tile, finished wood products, cement, and concrete) can often be left in place and cleaned if they are structurally sound, would be very difficult to replace, are only lightly contaminated on the surface, and can be successfully cleaned.
- Nonporous materials need to be thoroughly cleaned (includes metal, ceramic tile, porcelain, glass, hard plastics, highly finished solid wood items, and other hard smooth surfaces).

Step 3. Prepare parts of the work area, to minimize mold disturbance that will occur during containment set-up. Mold growth that such preparatory work would significantly disturb should be included in the containment area and cleaned after containment is erected.

- Remove easily accessible surface growth by HEPA vacuuming (a standard shop vacuum is not adequate) and damp wiping.
- Clean areas and materials by HEPA vacuuming or damp wiping, such as items that will be covered by critical barriers (e.g., air supply and return grilles); surfaces that will become inaccessible once containment is erected (e.g., flooring under containment wall); uncontaminated furniture and materials that will be removed from work area (if this can be done without agitating visible mold growth).

Step 4. Locally contain the affected area to minimize contaminant dispersal.

- Enclose areas of visible contamination and areas of suspected hidden growth with critical barriers (4–6 mm polyethylene, or comparable, sheeting of nonpermeable materials). The containment should be extended at least a few feet beyond areas of growth to ensure materials

Continued

Box 7.6 Practices and Procedures for "Moderate" Problem Remediation—cont'd

with heavy spore deposition are contained and to enable ease of re-mediation work.

- Critical barriers should block all openings so that mold particles cannot be carried outside the remediation area by air movement or through the mechanical ventilation system.

Step 5. Perform mold remediation activities, including detailed cleaning and/or removal of mold-contaminated materials.

- Porous materials: wrap or bag in plastic sheeting and discard in a secure disposal container. Clean surrounding nonporous materials at least two feet beyond visible growth. Securely bag waste and dispose.
- Semi-porous materials: remove, if necessary, or thoroughly clean as described for nonporous materials, disinfect, and dry. For stubborn problems, consider using a HEPA vacuum filtered sander, trisodium phosphate as a cleanser, or seal if the material can be kept dry.
- Nonporous materials: clean using a repeatable pattern of motions moving downward and from the cleanest areas to the dirtiest (not random washing or visually based) to ensure all surfaces have been thoroughly cleaned. Don't use methods such as sweeping, dry dusting or brushing. Perform cleaning in the following order.
  1. HEPA vacuum slowly and carefully.
  2. Damp wipe with a water and an all-purpose nonammonia based cleaner or detergent (work damp not wet).
  3. Once all surfaces have been dried from the initial cleaning, perform a second HEPA vacuuming in the opposite direction.
  4. Manage runoff and leave surfaces as dry as possible after cleaning.

Step 7. Decontaminate equipment and containment by thoroughly cleaning with a nonammonia based all-purpose cleaner followed by application of a mild bleach solution.

Step 8. Disassemble containment materials.

Step 9. Initiate additional drying if needed.

Step 10. Clean surrounding area as needed.

Step 11. Perform post-remediation evaluation to determine the effectiveness of remediation work and document findings.

Step 12. Periodically inspect for the presence of excess moisture and/or return of mold growth before rebuilding or refurnishing. If growth reappears, the moisture problem typically has not been properly addressed or corrected. Perform further investigation to determine moisture problem, correct the moisture problem, and remediate mold growth.

Step 13. Reconstruct and replace removed materials after moisture control has been achieved.

Box 7.7 Practices and Procedures for "Major" Problem Remediation

Step 1. Select appropriate personal protective equipment. Workers should protect themselves with the following:

- Respiratory protection capable of filtering particles down to 0.3 microns; a full-face HEPA filtered respirator, such as a P100 or powered air purifying respirator (PAPR), is strongly recommended
- Protective covering (disposable or washable outer clothing, long-sleeved tops, long pants)
- Eye protection (goggles that exclude fine dusts) if half-face HEPA respirator is used
- Anti-contamination garments
- Gloves (impervious to chemicals used and, if applicable, sharp objects)

Step 2. Determine if the material(s) supporting surface mold growth can be cleaned or should be removed and discarded.

- Porous materials (including drywall board, ceiling tile, insulation, unprotected "manufactured" or "processed" wood products, upholstered furniture, carpet, and padding) that are contaminated with mold need to be removed from the building. This should include all materials and furnishings that have, or had: visible mold growth; strong mold odors; or remained wet for longer than *48 hours* and are not easily cleanable.
- Hard-surfaced, semi-porous materials (e.g., tile, finished wood products, cement, and concrete) can often be left in place and cleaned, if they are structurally sound, would be very difficult to replace, are only lightly contaminated on the surface, and can be successfully cleaned.
- Nonporous materials need to be thoroughly cleaned (includes metal, ceramic tile, porcelain, glass, hard plastics, finished solid wood items, and other hard smooth surfaces).

Step 3. Prepare parts of work area to minimize mold disturbance that will occur during containment set-up. Mold growth that preparatory work would significantly disturb should be included in the containment area and cleaned after containment is erected.

- Remove easily accessible surface growth by HEPA vacuuming (a standard shop vacuum is not adequate) and damp wiping.
- Clean areas and materials by HEPA vacuuming or damp wiping, such as items that will be covered by critical barriers (e.g., air grilles); surfaces that will become inaccessible once the containment is erected (e.g., flooring under a containment wall); uncontaminated furniture and materials that will be removed from the work area (if this can be done without agitating the visible mold growth).

Step 4. Contain work area and limit access to authorized personnel.

- Erect containment around the area of visible and suspected hidden mold growth, extending several feet beyond the affected area. This should be designed to seal off the contaminated area in an airtight

Continued

Box 7.7 Practices and Procedures for "Major" Problem Remediation—cont'd

manner. An effective decontamination unit system should also be constructed for entering and exiting the remediation work area.

- Isolate the air handling system from work zone by sealing off supply and return grills with plastic sheeting and duct tape. If the area being remediated is served by an HVAC system, it should be shut down prior to any remedial activities.
- Use critical barriers (e.g., double layer of polyethylene and duct tape) to isolate the moldy area from clean occupied zones. Critical barriers should block all openings so that mold particles cannot be carried outside the remediation area by air movement or through the mechanical ventilation system.
- Establish a negative air-pressure differential of > 5 Pa or > 0.02 in. in a water column at all times between indoor areas external to the containment barriers and the enclosed remediation area. Establish negative pressure using HEPA-filtered ventilation equipment. Provide make-up air and test or monitor containment for leakage.

Step 5. Perform mold remediation activities, including detailed cleaning and/or removal of mold contaminated materials.

- Porous materials: wrap or bag in plastic sheeting and discard in a secure disposal container. Clean surrounding nonporous materials at least two feet beyond visible growth. Securely bag waste and dispose.
- Semi-porous materials: remove if necessary, or thoroughly clean as described for nonporous materials, disinfect, and dry. For stubborn problems, consider using HEPA vacuum filtered sander, trisodium phosphate as a cleanser, or seal if the material can be kept dry.
- Nonporous materials: clean using a repeatable pattern of motions moving downward and from the cleanest areas to the dirtiest (not random washing or visually based), to ensure all surfaces have been thoroughly cleaned. Don't use methods such as sweeping, dry dusting, or brushing). Perform cleaning in the following order.
  1. HEPA vacuum slowly and carefully.
  2. Damp wipe with water and an all-purpose nonammonia based cleaner or detergent (work damp not wet).
  3. Once all surfaces have been dried from the initial cleaning, perform a second HEPA vacuuming in the opposite direction.
  4. Manage runoff and leave surfaces as dry as possible after cleaning

Step 6. Decontaminate equipment and containment by thoroughly cleaning with a nonammonia based all-purpose cleaner followed by application of a mild bleach solution.

Step 7. Determine whether use of disinfectants is needed or desirable. Carefully follow the directions provided with the disinfectant. A dilute bleach solution may be adequate. The solution should be applied by light misting or wiping on (avoid runoff); treat the entire area that supported visible growth. The surfaces should be kept damp with the solution according to the manufacturer's recommendations. Allow to air dry. Wipe off residue.

Box 7.7 cont'd

Step 8. Clean surrounding area as needed.

Step 9. Allow or facilitate complete drying of all materials wet from excess moisture, cleaning activities, or disinfection solution. Dehumidifiers, fans, heat lamps, and ventilation with dry, warm air are among the methods that may be used to speed drying. Complete drying to normal levels may take days or weeks.

Step 10. Perform a thorough post-remediation evaluation and document work performed. Evaluate cleaning practices using previously identified evaluation methods to determine whether clearance requirements have been satisfied.

Step 11. Once post-remediation criteria have been achieved, deconstruct containment and then remove air-handling equipment. Air-handling equipment should be left running until entire containment has been completely taken down and removed.

Step 12. Periodically inspect for the continuing presence of excess moisture and/or return of mold growth before rebuilding or refurnishing. If growth reappears, the moisture problem has not been corrected. Perform further investigation to determine moisture problem, correct the moisture problem, and remediate mold growth.

Step 13. Reconstruct and replace removed materials after moisture control has been achieved.

Different states will likely have different rules, regulations, and recommendations. The material in this chapter may not be specific to your region. However, it does provide adequate guidance on the general principles of working with mold in a school environment. Always check with all agencies controlling work in your area before proceeding with work.

Licensing and Registration

8

The licensing and registration of mold remediation and abatement professionals is not a requirement in all states. However, some states have minimum standards that must be met for contractors who will be working with mold. You should check with your local jurisdiction to see if you are required to be certified, licensed, or registered to deal with mold in your region.

During my research for this book, I learned that Texas has comprehensive guidelines for mold contractors. For this reason, I use Texas as the model state for this chapter. Its regulations are not definitive of what other states may require, but they do offer a complete overview of the procedures you are likely to encounter if your state mandates control over mold contractors.[1] See Box 8.1 for licensing issues.

As we progress here you will be presented with rules and regulations based on specific elements. For example, you will see general conditions. This could be as simple as the rule that a person must be at least 18 years of age to participate as a mold contractor. Other sections will be more specific. I refer often to subchapters and specific paragraphs from the Texas Mold Assessment and Remediations Rules by number, with content shown in parentheses. Now that you know the ground rules, let's get busy.

► GENERAL CONDITIONS

A person must be licensed or registered in compliance with this subchapter to engage in mold assessment or mold remediation unless specifically exempted under §295.303 (relating to Exceptions and Exemptions). A person must be accredited as a mold training provider in compliance with this subchapter to offer mold training for fulfillment of specific training requirements for licensing under this subchapter. Each individual applying to be licensed or registered under this subchapter must be at least

---

[1] The complete Texas Mold Assessment and Remediation Rules document is available for download from *www.dshs.state.tx.us/mold/rules.shtm*.

> ### ▍Box 8.1 Licensing
>
> Where water damage has caused mold growth, homeowners and building managers may need to turn to outside contractors for inspection, assessment, and remediation. Currently, California has no requirements for licensing or certification of mold inspectors or remediators as such. Some other states have such requirements. The legislature might wish to evaluate whether similar licensing or certification programs are appropriate for California and might help homeowners and building managers to secure qualified assistance when needed to assess or remediate a mold problem.

18 years old at the time of application. A person licensed under this subchapter must maintain an office in Texas. An individual employed by a person licensed under this subchapter is considered to maintain an office in Texas through that employer.

An applicant for an initial license under §295.311 (relating to Mold Assessment Technician: Licensing Requirements), §295.312 (relating to Mold Assessment Consultant: Licensing Requirements), or §295.315 (relating to Mold Remediation Contractor: Licensing Requirements) must successfully complete an initial training course offered by a department-accredited training provider in that area of licensure and receive a course-completion certificate before applying for the license.

An applicant for renewal of a license listed under paragraph (1) of this subsection must successfully complete a refresher training course offered by a department-accredited training provider in the area of licensure for which renewal is sought and receive a course-completion certificate before applying for the renewal. The applicant must successfully complete the refresher course no later than 24 months after successful completion of the previous course.

An applicant for an initial or renewal registration under §295.314 (relating to Mold Remediation Worker: Registration Requirements) must successfully complete a training course as described under §295.320(d) and (f) (relating to Training: Required Mold Training Courses) and receive a course-completion certificate before applying for the registration. If a refresher course is required, the applicant must successfully complete the refresher course no later than 24 months after successful completion of the previous course.

In accordance with §295.310 (relating to Licensing: State Licensing Examination), an applicant for an initial license under §295.311, §295.312, or §295.315 must pass the state licensing examination in that area of licensure with a score of at least 70% correct before applying for the license. All applicants must pass the state examination within six months of completing any training course required under subsection (e)(1) of this section in three or fewer attempts or must successfully complete a new initial training course before retaking the state examination.

Each application for a credential or approval must provide all required information. An applicant shall indicate that a question does not apply by answering "not applicable" or "N/A." Applicants must submit complete applications, including all supporting documents, for each credential or approval sought.

An applicant for an initial license under §295.311, §295.312, or §295.315 must submit the complete application to the department within six months of passing the required state licensing examination, as evidenced by a postmark or shipping documents, or must successfully complete a new initial training course, receive a new training certificate, and pass a new state examination before submitting a new initial license application.

An applicant for an initial or renewal registration under §295.314 must submit the complete application to the department within 10 calendar days (not working days) of successfully completing the required training course, as evidenced by a postmark or shipping paperwork.

An applicant for a renewal of a license listed under paragraph (1) of this subsection must successfully complete a required refresher training course and receive a course-completion certificate before applying for renewal. The applicant must complete the refresher course no later than 24 months after completion of the previous course.

All credentials issued on or after January 1, 2006, are valid for two years and expire on the second anniversary of the effective date. Fees commensurate with a two-year credential must be included with any application for a credential that will expire on the second anniversary of its effective date. A credential holder is in violation of this subchapter if the holder practices with lapsed qualifications (see Box 8.2).

■ Box 8.2 General Responsibilities

Persons who are licensed, registered, or accredited under this subchapter shall:
- Adhere to the code of ethics prescribed by §295.304 (relating to Code of Ethics)
- Comply with work practices and procedures of this subchapter
- Refrain from engaging in activity prohibited under §295.307(a) (relating to Conflict of Interest and Disclosure Requirement)
- Maintain any insurance required under §295.309 (relating to Licensing: Insurance Requirements) while engaging in mold-related activities regulated under this subchapter
- Cooperate with department personnel in the discharge of their official duties, as described in §295.329 (relating to Compliance: Inspections and Investigations)
- Notify the department of changes in mailing address and telephone number

---

No credential, identification (ID) card, or approval issued under this subchapter shall be sold, assigned, or transferred. ID cards issued by the department must be present at the worksite any time an individual is engaged in mold-related activities. The department retains the right to confiscate and revoke any credential, ID card, or approval that has been altered. A mold assessment company, mold remediation company, mold analysis laboratory, or mold training provider that has been issued a credential under this subchapter:

- Shall designate one or more individuals as responsible persons; the credentialed person must notify the department in writing of any additions or deletions of responsible persons within 10 days of such occurrences
- Shall not transfer that credential to any other person, including to any company that has bought the credentialed entity; the credentialed entity must apply for a new credential within 60 days of being bought
- Must submit to the department a name-change application and a processing fee of $20 within 60 days of any change

All individuals who are required to be licensed or registered under this subchapter must have a valid department-issued ID card present at the worksite when engaged in mold-related activities, except as provided under §295.314(e) (relating to Mold Remediation Worker: Registration Requirements) for applicants for registration as mold remediation workers.

The license holder overseeing mold-related activities, with the exception of activities performed by a mold analysis laboratory, must ensure that a client and the property owner, if not the same, are provided a copy of the department's Consumer Mold Information Sheet prior to the initiation of any mold-related activity.

A credentialed person who becomes aware of violations of this subchapter must report these violations by the next business day to the department if, to that person's knowledge, the responsible party has not corrected the violations within that timeframe.

The individual designated by a licensed mold assessment company or mold remediation company as its responsible person shall not be the responsible person for another licensee with the same category of license. Credentialed persons are responsible for determining whether the mold-related activities in which they will engage require additional credentials beyond those required under this subchapter.

Conflict of Interest and Disclosure

A licensee shall not perform both mold assessment and mold remediation on the same project. A person shall not own an interest in an entity that performs mold assessment services and an entity that performs mold remediation services on the same project.

At the time of application for licensing, an applicant that is not an individual shall disclose to the department the name, address, and occupation of each person who has an ownership interest of 10% or more in the applicant. A licensee shall report to the department within 10 days any change related to a person who has an ownership interest of 10% or more including additions to or deletions from any list of such persons previously supplied to the department and any changes in the names, addresses, or occupations of any persons on such a list. This section does not apply to a license holder employed by a school district working on a project for that school district.

▶ APPLICATIONS AND RENEWAL FEES

Applications for a license, registration, or accreditation must be made on forms provided by the department and signed by the applicant. The department shall consider only complete applications. The application form must be accompanied by the following.

- A check or money order for the amount of the required fee made payable to the department, unless the application fee is paid through TexasOnline, as provided under the Texas Government Code, Chapter 2054, §2054.252 (relating to Texas-Online Project).
- For individuals applying for a credential, a current 1-inch square passport-quality color photograph of the applicant's face with a white background; a copy of the wallet-size photo-identification card from the applicable training course as required under §295.318(f)(6)(B) (relating to Mold Training Provider: Accreditation) must also be submitted.
- Proof that applicant meets all other requirements for obtaining the credential being sought.

Applicants who wish to discuss or obtain information concerning qualification requirements may contact the Department of State Health Services, Environmental and Sanitation Licensing Group. Applicants may visit the Mold Licensing Program's website at *www.dshs.state.tx.us/mold* to obtain information and download forms. The department may deny a credential to a person who fails to meet the standards established by this subchapter. Failure of the applicant to submit the required information and/or documentation within 90 days of issuance of a written notice of deficiency from the department will result in the application being denied.

The department shall refund application fees, less an administrative fee of $50 ($20 for remediation worker applications), if an applicant does not meet the requirements for the credential. The department shall refund fees paid in excess of the amounts required under this subchapter, less a $10 administrative fee. The department will not refund fees if the application was abandoned as a result of the applicant's failure to respond to a written request from the department for a period of 90 days. The applicant has the right to request a hearing in writing within 30 days of the date that is on the department's letter denying the credential. The hearing will be conducted in accordance with the Administrative Procedure Act (Texas Government Code, Chapter 2001) and the department's formal hearing rules.

At least 60 days before a person's license, registration, or accreditation is scheduled to expire, the department shall send a renewal notice by first-class mail to the person's last known address from the department's records. A person credentialed by the department retains full responsibility for supplying the department with a correct current address and phone number, and to take action to renew their credential whether or not they have received the notification from the department. The renewal notice will state the following.

• The type of credential requiring renewal
• The time period allowed for renewal
• The amount of the renewal fee
• How to obtain and submit a renewal application

A person seeking to renew a license, registration, or accreditation shall submit a renewal application no sooner than 60 days before the credential expires. The department shall renew the license, registration, or accreditation for a term as provided under §295.305(h) (relating to Credentials: General Conditions) if the person:

• Is qualified to be credentialed
• Pays to the department the nonrefundable renewal fee
• Submits to the department a renewal application on the prescribed form along with all required documentation
• Has complied with all final orders resulting from any violations of this subchapter, unless an exception is granted in writing by the department and submitted with the application.

A person shall not perform any mold-related activity with an expired license, registration, or accreditation. If a person makes a timely and complete application for the renewal of a valid credential, the credential does not expire until the department has finally granted or denied the application. The department shall renew a credential that has been expired for 180 days or less if the person meets the requirements of subsection (f) of this section. A person whose credential has been expired for more than 180 days must obtain a new credential and must comply with current requirements and procedures, including any state examination requirements. A person desiring a replacement credential or ID card shall submit a request in writing on a department-issued form, with a $20 fee.

▶ INSURANCE REQUIREMENTS

Persons required to have insurance must, at a minimum, obtain policies for commercial general liability in the amount of not less than $1 million per occurrence. Governmental entities that are self-insured are not required to purchase insurance under this subchapter. A nongovernmental entity (business entity or individual) may be self-insured if it submits to the department for approval an affidavit signed by an authorized official of the entity or by the individual stating that it has a net worth of at least $1 million. A current financial statement indicating a net worth of at least $1 million must accompany the affidavit. A new affidavit and current financial statement must be submitted with each renewal application.

An individual required to have insurance must obtain individual coverage unless covered under the policy of the individual's employer or employed by a governmental entity or a person approved by the department to be self-insured. Insurance policies required under this section must be currently in force and must be written by one of the following:

- An insurance company authorized to do business in Texas
- An eligible Texas surplus lines insurer as defined in Texas Insurance Code, Article 1.14-2 (relating to Surplus Lines Insurance)
- A Texas registered risk retention group
- A Texas registered purchasing group

The certificate of insurance must be complete, including all of the applicable coverages and endorsements, and must name the Department of State Health Services, Environmental and Sanitation Licensing Group as a certificate holder. Each required policy shall be endorsed to provide the department with at least a 10-day notice of cancellation or material change for any reason. An applicant for an initial or renewal license must provide proof of insurance in one of the following forms:

- A copy of the required current certificate of insurance
- If claiming to be self-insured, a statement that it is a governmental entity, or, if a nongovernmental entity, the affidavit and current financial statement described under subsection (a) of this section
- Proof that the applicant is employed by a licensed mold assessment or remediation company that has the required insurance

The department may impose an administrative penalty or take other disciplinary action against any person who fails to have the current insurance required under this section.

If a policy is canceled or materially changed, the licensee shall notify the department in writing not later than 20 calendar days prior to the change or cancellation effective date. A licensed company may file a single notification for the company and its licensed employees. If a policy expires or is canceled or materially changed, the licensee shall cease work. Prior to resuming work, the licensee must do one of the following:

• Provide to the department a certificate of the renewal or replacement policy
• Submit to the department the affidavit and current financial statement described under subsection (a) of this section and receive departmental approval to be self-insured

If an individual licensee ceases to be covered under an employer's insurance, the individual must obtain replacement coverage either individually or through a new employer. The individual must submit the documentation required under subsection (c) of this section to the department before engaging in any mold-related activities.

▶ EXAMINATION AND LICENSING REQUIREMENTS

An applicant for an initial individual license who has successfully completed the required training course from a department-accredited training provider must pass the state examination with a score of at least 70% correct prior to applying for the license. Applicant must pass the examination within six months of completing the training course. See Box 8.3 for information on OSHA requirements.

An individual is permitted to take two reexaminations after failing an initial examination. An individual who fails both reexaminations must repeat the initial training course, submit a new application for the state examination, and provide a copy of the new training certificate.

Annually, the department shall publish a schedule of examination dates and locations. Training providers shall provide state examination schedules as a part of their instruction. Registrations must be submitted by mailing, faxing, or emailing a registration form to the Department of State Health Services, Environmental

> **Box 8.3 Construction Workers and Maintenance and Custodial Personnel**
>
> Construction workers, including those who are doing home and commercial building remodeling and repair, are subject to OSHA and Cal/OSHA hazard communication rules. Although the regulations are designed primarily with a view to chemicals and other manufactured substances, the principle of hazard communication could be extended to include risks from naturally occurring contamination, such as molds.
>
> The "safe and healthful workplace" standard could be understood to extend to those risks. Similar considerations to those for construction workers apply to maintenance and custodial personnel whose work exposes them to such biological contaminants as bacteria, viruses, fungi, dust mites, and cockroaches.

and Sanitation Licensing Group and must be received by the department no later than five working days before the examination date. Information on the examination schedule and assistance with registration is available by contacting the Department of State Health Services, Environmental and Sanitation Licensing Group. Entrance into the examination site will be allowed only on presentation of a valid photo identification from an accredited training provider. Companies with 30 or more employees to be tested may call the department to arrange an additional examination date for a $50 per person examination fee.

A fee of $25 is required for any examination or reexamination. A fee of $50 per person shall be paid for examinations administered at locations and times other than those published. The department must receive the required fees no later than five working days before the examination. A grade of at least 70% correct must be achieved to pass the examination. Scores will be reported only by mail no later than 30 working days after the date the examination is taken. Information regarding reexamination, if necessary, will be included. If requested in writing by an individual who fails a licensing examination, the department shall furnish the individual with an analysis of the individual's performance on the examination.

Mold Assessment Technician License

Unless exempted under §295.303 (relating to Exceptions and Exemptions), an individual must be licensed as a mold assessment technician to perform activities listed under subsection (b) of this section, except that an individual licensed under §295.312 (relating to Mold Assessment Consultant: Licensing Requirements) is not required to be separately licensed under this section.

An individual licensed under this section is authorized to determine the location and extent of mold or suspected mold that is present in a facility. A mold assessment technician is licensed to:

- Record visual observations and take onsite measurements, including temperature, humidity, and moisture levels, during an initial or post-remediation mold assessment
- Collect various samples for analysis during an initial mold assessment
- Prepare a mold assessment report
- As directed by an onsite assessment consultant, collect samples during a post-remediation mold assessment

In addition to the requirements for all applicants who are listed in §295.305 (relating to Credentials: General Conditions) and §295.309 (relating to Licensing: Insurance Requirements), an applicant must be a high school graduate or have obtained a General Educational Development (GED) certificate. The fees for a mold assessment technician license are:

- $200 for the license
- A required Texas Online subscription and convenience fee

Applications shall be submitted as required by §295.308(a) (relating to Credentials: Applications and Renewals). If the application is for an initial license, an applicant shall include the following:

- A copy of a high school diploma or GED certificate
- Proof of compliance with the insurance requirement specified in §295.309
- A copy of a certificate of training as described in §295.320(b) (relating to Training: Required Mold Training Courses)
- Proof of successfully passing the state licensing examination with a score of at least 70% correct

If the application is for renewal of a license, an applicant shall include the following:

- A copy of a certificate of training as described in §295.320(g), unless the applicant is exempt under §295.305(g)(3)
- Proof of compliance with the insurance requirement specified in §295.309

In addition to the requirements listed in §295.306 (relating to Credentials: General Responsibilities), a licensed mold assessment technician shall:

- Perform only activities allowed under subsection (b) of this section
- Comply with mold-sampling protocols accepted as industry standards, as presented in the training course materials or as required by his and/or her employer
- Utilize the services of a laboratory that is licensed by the department to provide analysis of mold samples
- Provide to the client a mold assessment report following an initial (preremediation) assessment for mold, if the technician is not acting as an employee of a licensed mold assessment consultant or company

Mold Assessment Consultant License

Unless exempted under §295.303 (relating to Exceptions and Exemptions), an individual must be licensed as a mold assessment consultant to perform activities listed under subsection (b) of this section. A licensed mold assessment consultant who employs two or more individuals required to be licensed under this section or §295.311 (relating to Mold Assessment Technician: Licensing Requirements) must be separately licensed as a mold assessment company under §295.313 (relating to Mold Assessment Company: Licensing Requirements), except that an individual licensed as a mold assessment consultant and doing business as a sole proprietorship is not required to be separately licensed under §295.313.

An individual licensed under this section is also licensed to perform all activities of a mold assessment technician listed in

§295.311(b) and (f). In addition, a licensed mold assessment consultant is licensed to:

- Plan surveys to identify conditions favorable for indoor mold growth or to determine the presence, extent, amount, or identity of mold or suspected mold in a building
- Conduct activities recommended in a plan developed under paragraph (1) of this subsection and describe and interpret the results of those activities
- Determine locations at which a licensed mold assessment technician will record observations, take measurements, or collect samples
- Prepare a mold assessment report, including the observations made, measurements taken, and locations and analysis results of samples taken by the consultant or by a licensed mold assessment technician during the mold assessment
- Develop a mold management plan for a building, including recommendations for periodic surveillance, response actions, and prevention and control of mold growth
- Prepare a mold remediation protocol, including the evaluation and selection of appropriate methods, personal protective equipment, engineering controls, project layout, post-remediation clearance evaluation methods and criteria, and preparation of plans and specifications
- Evaluate a mold remediation project for the purpose of certifying that mold contamination identified for the remediation project has been remediated as outlined in a mold remediation protocol
- Evaluate a mold remediation project for the purpose of certifying that the underlying cause of the mold has been remediated so that it is reasonably certain that the mold will not return from that remediated cause
- Complete appropriate sections of a Certificate of Mold Damage Remediation as specified under §295.327(b) (relating to Photographs; Certificate of Mold Damage Remediation; Duty of Property Owner)

In addition to the requirements for all applicants listed in §295.305 (relating to Credentials: General Conditions) and §295.309 (relating to Licensing: Insurance Requirements), an applicant must meet at least one of the following education and/ or experience requirements:

- A bachelor's degree from an accredited college or university with a major in a natural or physical science, engineering, architecture, building construction, or building sciences, and at least one year of experience in an allied field
- At least 60 college credit hours with a grade of C or better in the natural sciences, physical sciences, environmental sciences, building sciences, or a field related to any of those sciences, and at least three years of experience in an allied field
- A high school diploma or a General Educational Development (GED) certificate and at least five years of experience in an allied field
- Certification as an industrial hygienist, a professional engineer, a professional registered sanitarian, a certified safety professional, or a registered architect, with at least one year of experience in an allied field

The fees for a mold assessment consultant license are:

- $600 for the license
- A required Texas Online subscription and convenience fee

Applications shall be submitted as required by §295.308(a) (relating to Credentials: Applications and Renewals). If the application is for an initial license, an applicant shall include the following in the application package:

- Verifiable evidence that the applicant meets at least one of the eligibility requirements under subsection (c)(1) through (4) of this section
- Proof of compliance with the insurance requirement specified in §295.309
- Proof of successfully passing the state licensing examination with a score of at least 70% correct
- A copy of a certificate of training as described in §295.320(c) (relating to Training: Required Mold Training Courses)

If the application is for renewal of a license, an applicant shall include the following in the application package:

- A copy of a certificate of training as described in §295.320(g), unless the applicant is exempt under §295.305(g)(3)

- Proof of compliance with the insurance requirement specified in §295.309

In addition to the requirements listed in §295.306 (relating to Credentials: General Responsibilities), a licensed mold assessment consultant shall:

- Provide adequate consultation to the client to diminish or eliminate hazards or potential hazards to building occupants caused by the presence of mold growth in buildings
- Provide, in accordance with a client's instructions, professional services concerning surveys, building conditions that have or might have contributed to mold growth, proper building operations and maintenance to prevent mold growth, and compliance with work practices and standards
- Comply with mold sampling protocols as presented in training course materials or as required by his/her employer
- Inquire of the client whether any hazardous materials, including lead-based paint and asbestos, are present in the project area
- Ensure that all employees who will conduct mold assessment activities are provided with, fit-tested for, and trained in the correct use of personal protective equipment appropriate for the activities to be performed
- Ensure that the training and license of each licensed employee are current, as described in §295.320 and §295.311 or §295.312, respectively
- Provide to the client a mold assessment report following an initial (pre-remediation) mold assessment; if the consultant includes the results of the initial assessment in a mold remediation protocol or a mold management plan, a separate assessment report is not required
- Provide to the client a mold remediation protocol before a remediation project begins
- Utilize the services of a laboratory that is licensed by the department to provide analysis of mold samples
- If he/she performs post-remediation assessment on a project and ceases to be involved with the project before it achieves clearance, provide a final status report to the client and to the mold remediation contractor or company performing mold remediation work for the client as specified under §295.324(e) (relating to Post-Remediation Assessment and Clearance)

- Provide a passed clearance report to the client as specified under §295.324(d) and complete applicable sections of a Certificate of Mold Damage Remediation as specified under §295.327(b) (relating to sections on Photographs, Certificate of Mold Damage Remediation, Duty of Property Owner)
- Comply with recordkeeping responsibilities under §295.326(c) (relating to Recordkeeping)
- Sign and date each mold assessment report and each mold management plan that he/she prepares and include his/her license number and expiration date on each report and each plan
- Sign and date each mold remediation protocol on the cover page, including his/her license number and expiration date; the consultant must also initial the protocol on every page that addresses the scope of work and on all drawings related to the remediation work
- Review and approve changes to any protocol by signing or initialing according to paragraph (14) of this subsection

Mold Assessment Company Licensing

A person performing mold assessment work on or after January 1, 2005 must be licensed as a mold assessment company if the person employs two or more individuals required to be licensed under §295.311 (relating to Mold Assessment Technician: Licensing Requirements) or §295.312 (relating to Mold Assessment Consultant: Licensing Requirements), except that an individual licensed as a mold assessment consultant and doing business as a sole proprietorship is not required to be separately licensed under this section. A mold assessment company shall designate one or more individuals licensed as mold assessment consultants as its responsible person(s).

As a condition of licensure, a mold assessment company must:

- Notify the department in writing of any changes in individual licensed mold assessment consultants as responsible persons within 10 days of such occurrences
- Maintain commercial general liability insurance, as described in §295.309 (relating to Licensing: Insurance Requirements)

- Refrain from mold assessment activity during any period without the active employment of at least one individual licensed mold assessment consultant designated as the responsible person for the company
- Notify the department in writing of any change related to a person who has an ownership interest of 10% or more (including additions to or deletions from any list of such persons previously supplied to the department and any changes in the names, addresses, or occupations of any persons on such a list) within 10 days of the change
- Refrain from engaging in activity prohibited under §295.307(a) (relating to Conflict of Interest and Disclosure Requirement)

To be eligible for licensing, an applicant must:

- Employ at least one licensed mold assessment consultant
- Maintain an office in Texas

The fees for a mold assessment company license are $1000 for the license and a required Texas Online subscription and convenience fee. Applications shall be submitted as required by §295.308(a) (relating to Credentials: Applications and Renewals). An applicant shall include the following in the application package:

- Proof of compliance with the insurance requirement specified in §295.309
- The name, address, and occupation of each person who has an ownership interest of 10% or more in the company
- The name and license number of each licensed mold assessment consultant designated by the applicant as a responsible person

In addition to the requirements as listed in §295.306 (relating to Credentials: General Responsibilities), a licensed mold assessment company shall:

- Follow the recordkeeping requirements, at both the Texas office and work site locations, as described in §295.326(c) (relating to Recordkeeping)
- Provide each client with a mold assessment report following an initial (preremediation) mold assessment; if the company includes the results of the initial assessment in a mold remediation protocol or a mold management plan, a separate assessment report is not required

- Provide each client a mold remediation protocol before remediation begins
- Ensure that all employees who will conduct mold assessment activities are provided with, fit tested for, and trained in the correct use of personal protective equipment appropriate for the activities to be performed
- Ensure that the training and license of each licensed employee are current, as described in §295.320 (relating to Training: Required Mold Training Courses) and §295.311 or §295.312, respectively
- Utilize the services of a laboratory that is licensed by the department to provide analysis of mold samples
- Maintain commercial general liability insurance, as described in §295.309
- If the company performs post-remediation assessment on a project and ceases to be involved with the project before it achieves clearance, provide a final status report to the client and to the mold remediation contractor or company performing mold remediation work for the client as specified under §295.324(e) (relating to Post-Remediation Assessment and Clearance)
- Provide a passed clearance report to the client as specified under §295.324(d) and provide a Certificate of Mold Damage Remediation, with applicable sections completed by a mold assessment consultant, to a mold remediation company or contractor, as specified under §295.327(b) (relating to Photographs, Certificate of Mold Damage Remediation, Duty of Property Owner)

Mold Remediation Worker Registration

Unless exempted under §295.303 (relating to Exceptions and Exemptions), an individual must be registered as a mold remediation worker to perform mold remediation, except that an individual licensed under §295.315 (relating to Mold Remediation Contractor: Licensing Requirements) is not required to be separately registered under this section.

In addition to the requirements for all applicants listed in §295.305 (relating to Credentials: General Conditions), an applicant must:

- Be employed by a licensed mold remediation contractor or company

- Complete a mold remediation worker training course provided by either the applicant's employer or by an accredited mold training provider, as described under §295.320(d) (relating to Training: Required Mold Training Courses)

The fees for a mold remediation worker registration are:

- $60 for the license
- A required Texas Online subscription and convenience fee

Applications shall be submitted as required by §295.308(a) (relating to Credentials: Applications and Renewals) and shall include a copy of the training certificate required under §295.320(d)(5)(A), unless the applicant is exempt under §295.305(g)(3). An applicant must submit an application to the department within 10 calendar days of completing a worker training course, as evidenced by a postmark or shipping paperwork.

An individual who has successfully completed remediation worker training and received a training certificate may perform mold remediation work allowed under this section for a period of not more than 30 days from the training date if:

- The individual has submitted an application for registration to the department as required under subsection (d) of this section
- A copy of the training certificate is present at the work site at all times while the individual engages in mold remediation
- The individual is in possession of a valid government-issued photo identification at all times while performing mold remediation work

In addition to the requirements as listed in §295.306 (relating to Credentials: General Responsibilities), a registered mold remediation worker shall use remediation techniques specified in the project mold remediation work plan. Registered mold remediation workers are prohibited from:

- Performing mold remediation except under the supervision, as defined in §295.303(f), of a licensed remediation contractor
- Engaging in any mold-related activity requiring licensing as a remediation contractor under this subchapter

Mold Remediation Contractor Licensing

Unless exempted under §295.303 (relating to Exceptions and Exemptions), an individual must be licensed as a mold remediation contractor to perform activities listed under subsection (b) of this section. A licensed mold remediation contractor who employs one or more individuals required to be licensed under this section or §295.314 (relating to Mold Remediation Worker: Registration Requirements) must be separately licensed as a mold remediation company under §295.316 (relating to Mold Remediation Company: Licensing Requirements), except that an individual licensed as a mold remediation contractor and doing business as a sole proprietorship is not required to be separately licensed under §295.316.

An individual licensed under this section may perform mold remediation and supervise registered mold remediation workers performing mold remediation. In addition, a licensed mold remediation contractor is licensed to provide mold remediation services including:

- Preparing a mold remediation work plan providing instructions for the remediation efforts to be performed for a mold remediation project
- Conducting and interpreting the results of activities recommended in a work plan developed under paragraph (1) of this subsection, including any of the activities of a registered mold remediation worker under §295.314

In addition to the requirements for all applicants listed in §295.305 (relating to Credentials: General Conditions) and §295.309 (relating to Licensing: Insurance Requirements), an applicant must meet at least one of the following education and/or experience requirements:

- A bachelor's degree from an accredited college or university with a major in a natural or physical science, engineering, architecture, building construction, or building sciences and at least one year of experience either in an allied field or as a general contractor in building construction
- At least 60 college credit hours with a grade of C or better in the natural sciences, physical sciences, environmental sciences,

building sciences, or a field related to any of those sciences, and at least three years of experience in an allied field or as a general contractor in building construction

- A high school diploma or GED certificate, plus at least five years of experience in an allied field or as a general contractor in building construction
- Certification as an industrial hygienist, a professional engineer, a professional registered sanitarian, a certified safety professional, or a registered architect, with at least one year of experience either in an allied field or as a general contractor in building construction

The fees for a mold remediation contractor license are $500 for the license; and a required Texas Online subscription and convenience fee.

Applications shall be submitted as required by §295.308(a) (relating to Credentials: Applications and Renewals). If the application is for an initial license, an applicant shall include the following in the application package:

- Verifiable evidence that the applicant meets at least one of the qualifications under subsection (c)(1) of this section
- Proof of compliance with the insurance requirement specified in §295.309
- A copy of a certificate of training indicating successful completion within the past 12 months of an initial training course offered by a department-accredited training provider as described in §295.320(e) (relating to Training: Required Mold Training Courses)
- Proof of successfully passing the state licensing examination with a score of at least 70% correct

If the application is for renewal of a license, an applicant shall include the following in the application package:

- A copy of a certificate of training as described in §295.320(g), unless the applicant is exempt under §295.305(g)(3)
- Proof of compliance with the insurance requirement specified in §295.309

In addition to the requirements as listed in §295.306 (relating to Credentials: General Responsibilities), the mold remediation contractor shall be responsible for:

- Supervising mold remediation workers as defined in Section 295.302(41) (relating to Definitions)
- Accurately interpreting field notes, drawings, and reports relating to mold assessments
- Advising clients about options for mold remediation
- Complying with standards for preparing mold remediation work plans, as presented in training course materials or as required by the mold remediation company by whom the contractor is employed
- Providing to a client a mold remediation work plan for the project before the mold remediation preparation work begins
- Inquiring of the client whether or not any known or suspected hazardous materials, including lead-based paint and asbestos, are present in the project area
- Signing and dating each mold remediation work plan that he or she prepares on the cover page; the cover page shall also include his or her license number and expiration date; he or she must also initial the work plan on every page that addresses the scope of work and on all drawings related to the remediation work
- Submitting the required notification to the department, as described in §295.325 (relating to Notifications), unless employed by a licensed mold remediation company
- Ensuring that all individuals are provided with, fit-tested for, and trained in the correct use of personal protective equipment required under §295.322(c) (relating to Minimum Work Practices and Procedures for Mold Remediation)

If the mold remediation contractor is doing business as a sole proprietorship and is not required to be separately licensed as a mold remediation company under §295.316, he/she shall be responsible for:

- Ensuring that the training, as described in §295.320 (relating to Training: Required Mold Training Courses), and license of each employee who is required to be licensed under this sub-chapter is current

- Ensuring that the training, as described in §295.320, and registration of each registered employee is current
- Ensuring that each unregistered employee who is required to be registered under this subchapter is provided the training required under §295.320(d) before performing any mold remediation work
- Complying with all requirements under §295.320(d) if the contractor provides the training
- Ensuring that a previously unregistered employee who is provided training as specified in subparagraph (c) of this paragraph:
  - Has applied to the department for registration before allowing that employee to perform any mold remediation work, except as provided under §295.314(e)
  - Is registered before allowing that employee to perform any mold remediation work more than 30 days after the date of the training, in accordance with §295.314(e)
- Complies with recordkeeping responsibilities under §295.326 (relating to Recordkeeping)
- Provides to the property owner a completed Certificate of Mold Damage Remediation as specified under §295.327 (relating to Photographs, Certificate of Mold Damage Remediation, Duty of Property Owner)

Mold Remediation Company Licensing

A person performing mold remediation work must be licensed as a mold remediation company if the person employs one or more individuals required to be registered under §295.314 (relating to Mold Remediation Worker: Registration Requirements) or licensed under §295.315 (relating to Mold Remediation Contractor: Licensing Requirements), except that an individual licensed as a mold remediation contractor and doing business as a sole proprietorship is not required to be separately licensed under this section. A mold remediation company shall designate one or more individuals licensed as mold remediation contractors as its responsible person(s).

A licensed mold remediation company is specifically authorized to employ remediation contractors and remediation workers who are currently licensed or registered under this subchapter to assist in the company's mold remediation activity. As a condition of licensure, a mold remediation company must:

- Employ at least one licensed remediation contractor and refrain from remediation activity during any period without the active employment of at least one individual licensed mold remediation contractor designated as the responsible person for the company
- Notify the department in writing of any additions or deletions of responsible persons within 10 days of such occurrences
- Maintain commercial general liability insurance, as described under §295.309 (relating to Licensing: Insurance Requirements)
- Notify the department in writing of any change related to a person who has an ownership interest of 10% or more (including additions to or deletions from any list of such persons previously supplied to the department and any changes in the names, addresses, or occupations of any persons on such a list) within 10 days of the change
- Refrain from engaging in activity prohibited under §295.307(a) (relating to Conflict of Interest and Disclosure Requirement)

The fees for a mold remediation company license are $1000 for the license; and a required Texas Online subscription and convenience fee.

Applications shall be submitted as required by §295.308(a) (relating to Credentials: Applications and Renewals). An applicant shall include the following in the application package:

- Proof of compliance with the insurance requirement specified in §295.309
- The name, address, and occupation of each person who has an ownership interest of 10% or more in the company
- The name and license number of each licensed mold remediation contractor designated by the applicant as a responsible person

In addition to the requirements as listed in §295.306 (relating to Credentials: General Responsibilities), the mold remediation company shall be responsible for:

- Complying with recordkeeping requirements, at both central office and work site locations, as described in §295.326 (relating to Recordkeeping)

- Submitting the required notification to the department, as required under §295.325 (relating to Notifications)
- Providing to each client a mold remediation work plan for the project before the mold remediation preparation work begins
- Ensuring that all employees who will conduct mold remediation activities are provided with, fit-tested for, and trained in the correct use of personal protective equipment required under §295.322 (relating to Minimum Work Practices and Procedures for Mold Remediation)
- Ensuring that the training, as described in §295.320 (relating to Training: Required Mold Training Courses), and license of each employee who is required to be licensed under this subchapter is current
- Ensuring that the training, as described in §295.320, and registration of each registered employee is current
- Ensuring that each unregistered employee who is required to be registered under this subchapter is provided the training required under §295.320(d) before performing any mold remediation work
- Complying with all requirements under §295.320(d) if the company provides the training
- Ensuring that a previously unregistered employee who is provided training as specified in paragraph (7) of this paragraph:
  - Has applied to the department for registration before allowing that employee to perform any mold remediation work, except as provided under §295.314(e)
  - Is registered before allowing that employee to perform any mold remediation work more than 30 days after the date of the training, in accordance with §295.314(e)

▶ MOLD TRAINING REQUIREMENTS

Individual applicants for licensing or renewal must submit evidence acceptable to the department of fulfillment of specific training requirements.

The assessment technician course shall consist of at least 24 training hours that includes lectures, demonstrations, audio-visuals and hands-on training, course review, and a written test of 50 multiple-choice questions. The assessment technician course shall include:

- Sources of, conditions necessary for, and prevention of indoor mold growth

- Potential health effects, in accordance with a training protocol developed in consultation with state professional associations, including at least one representing physicians
- Workplace hazards and safety, including personal protective equipment, and respirators
- Technical and legal considerations for mold assessment, including applicable regulatory requirements, the role of the mold assessment technician, and the roles of other professionals (including an assessment consultant)
- Performance of visual inspections where mold might be present and the determination of sources of moisture problems, including exterior spaces (e.g., crawlspaces and attics), interior components (e.g., windows, plumbing, walls, and ceilings), and heating, ventilation, and air-conditioning (HVAC) systems (e.g., return air and supply ducts)
- Utilization of physical measurement equipment and tools, including moisture meters, humidity meters, particle counters, data-logging equipment, and visual and robotic inspection equipment
- Biological sampling strategies and methodologies, including sampling locations and techniques, and minimizing cross-contamination
- Sampling methodologies, including bulk, surface (including tape, swab, and vacuum sampling), and air sampling (including the differences between culturable and particulate sampling, sampling times, calibrating pumps, selecting media for culturable samples, and sampling for fungal volatile organic compounds)
- State-of-the-art work practices and new technologies
- Proper documentation for reports, including field notes, measurement data, photographs, structural diagrams, and chain-of-custody forms
- An overview of mold remediation projects and requirements, including containment and air filtration
- Clearance testing and procedures, including review of mold remediation protocols, work plans, visual inspections, and sampling strategies

The assessment consultant course shall consist of at least 40 training hours that needs to include lectures, demonstrations, audio-visuals and hands-on training, course review, and a written test of 100 multiple-choice questions. The assessment consultant course shall include:

- All topics listed under subsection (b) of this section, including appropriate hands-on activities
- Requirements concerning workplace safety, including components of and development of respiratory protection plans and programs, workplace safety plans, and medical surveillance programs
- Technical and legal considerations for mold assessment, including applicable regulatory requirements, the role of the assessment consultant, the roles of other professionals, record-keeping and notification requirements, insurance, and legal liabilities
- An overview of building construction, building sciences, moisture control, and water intrusion events
- Prevention of indoor air quality problems, including avoiding design and construction defects and improving maintenance and housekeeping
- Basics of HVAC systems and their relationship to IAQ (including pyschrometrics, filtration, ventilation, and humidity control), HVAC inspection and assessment, and remediation of HVAC systems
- Survey protocols for effective assessment, covering the areas described under subsections (b)(5) through (b)(8) of this section
- Interpretation of data and sampling results
- Interviewing building occupants, minimum requirements for questionnaires, and interpreting results
- Writing mold management plans and mold remediation protocols, including format and contents (including structural components, HVAC systems, and building contents), defining affected areas (including floor plans), identifying and repairing moisture sources and their causes, developing a scope of work analysis, specifying containment and air filtration strategies, determining post-remediation assessment criteria, and clearance criteria
- Post-remediation clearance testing and procedures, including review of mold remediation plans, visual inspections, sampling strategies, quality assurance, and case studies

Remediation worker training shall consist of at least four training hours that includes lectures, demonstrations, audio-visuals, and hands-on training. The training shall include all course information and material required under this subsection. An individual must successfully complete worker training and submit an application for registration as a mold remediation worker prior to performing any mold remediation. The training must be provided by either the licensed mold remediation contractor or company employing the individual receiving the training or a mold training provider accredited by the department. The principal instructor for the training must be either a licensed mold remediation contractor or an individual who is approved by the department under §295.319 to teach mold-related courses.

The training shall adequately address the following areas and shall include hands-on training.

- Sources of indoor mold and conditions necessary for indoor mold growth
- Potential health effects and symptoms from mold exposure, in accordance with a training protocol developed in consultation with state professional associations, including at least one representing physicians
- Workplace hazards and safety, personal protective equipment including respirators, personal hygiene, personal decontamination, confined spaces, and water, structural, and electrical hazards
- Technical and legal considerations for mold remediation, including applicable regulatory requirements, the role of the worker, and the roles of other professionals
- An overview of how mold remediation projects are conducted, including containment and air filtration
- Work practices for removing, cleaning, and treating mold

The person providing the training shall submit to the department, within five working days of a training session, the following items, on a form provided by the department:

- The name, address, telephone number, and license number of the person listed under paragraph (1) of this subsection who provided the training

- The date of the training
- The printed name, address, telephone number, number identifier, and signature of each individual who attended the training
- The printed name and signature of the principal instructor
- A group color photograph, taken at the end of the training, that identifies each individual who attended the training (digital or scanned images will be accepted if persons are easily identifiable); the group color photograph must be no smaller than a standard 3.5-inch by 4.25-inch print
- A statement indicating which individuals successfully completed the training and which individuals did not

The person providing the training shall provide a training certificate to each individual who successfully completes the training. Each certificate must include:

- The name, address, telephone number, and license number of the person listed under paragraph (1) of this subsection who provided the training
- The date of the training
- The name, address, telephone number, and number identifier of the individual
- The printed name and signature of the principal instructor
- A statement that the individual successfully completed training
- A current 1-inch square passport-quality color photograph of the individual's face on a white background, taken during the course, to be attached by the individual to an application for registration
- A copy of the registration application

The person providing the training must maintain a file for each training session that includes the date, the certificate numbers, and the names, addresses, and telephone numbers of students receiving training certificates. All information from the training must correspond to the information on each certificate.

The remediation contractor course shall consist of at least 40 training hours that includes lectures, demonstrations, audiovisuals and hands-on training, course review, and a written test of 100 multiple-choice questions. The course requirements in paragraphs (3), (7), and (8) of this subsection require hands-on training as an integral part of the training. The course shall adequately address:

- Sources of indoor mold and conditions necessary for indoor mold growth
- Potential health effects, in accordance with a training protocol developed in consultation with state professional associations, including at least one representing physicians
- Requirements concerning workplace hazards and safety, personal protective equipment including respirators, personal hygiene, personal decontamination, confined spaces, and water, structural, and electrical hazards
- Requirements concerning worker protection, which should include components of and development of respiratory protection plans and programs, workplace safety plans, and medical surveillance programs
- Technical and legal considerations for mold remediation, including applicable regulatory requirements, the role of the mold remediation contractor, the role of the mold remediation worker, the roles of other professionals, insurance, legal liabilities, and recordkeeping and notification requirements
- Building sciences, moisture control, and water intrusion events
- An overview of how mold remediation projects are conducted and requirements thereof, including containment and air filtration
- Work practices for removing, cleaning, and treating mold, including state-of-the-art work practices and new technologies
- Development of a mold remediation work plan from a protocol, including writing the work plan, detailing remediation techniques for the building structure, HVAC system, and contents, delineating affected areas from floor plans, developing appropriate containment designs, determining HEPA air-filtration requirements, and determining dehumidification requirements
- Clearance testing and procedures, including a review of typical clearance criteria, visual inspection of the work area prior to clearance, and achieving clearance
- Contract specifications, including estimating job costs from a protocol and determining insurance and liability issues
- Protecting the public and building occupants from mold exposure

The refresher courses for mold assessment technicians, mold assessment consultants, and mold remediation contractors shall be at least eight training hours in length. Refresher training for mold remediation workers shall be at least four (4) training hours in length and shall be provided by a person specified under subsection (d)(1) of this section. Refresher training shall include a review of state regulations, state-of-the-art developments, and key aspects of the initial training course. All individual licensees and registrants shall receive refresher training every two years.

Each training provider shall administer a closed-book written test to students who have completed an initial or refresher training course, except that no examination is required of students in remediation worker training. The initial test for assessment technician training shall consist of 50 multiple-choice questions, and the initial tests for assessment consultant training and remediation contractor training shall consist of 100 multiple-choice questions. Training providers may include demonstration testing as part of the initial test. The refresher tests shall consist of at least 10 questions. A student must answer correctly at least 70% of the questions to receive a course-completion certificate. Training providers shall use tests provided or approved by the department.

▶ MINIMUM WORK PRACTICES AND PROCEDURES

These general work practices are minimum requirements and do not constitute complete or sufficient specifications for mold assessment. More detailed requirements developed by an assessment consultant for a particular mold remediation project shall take precedence over the provisions of this section.

Mold Assessment

The purpose of a mold assessment is to determine the sources, locations, and extent of mold growth in a building; to determine the condition(s) that caused the mold growth; and to enable the assessment consultant to prepare a mold remediation protocol. If an assessment consultant or company determines that personal protective equipment (PPE) should be used during a mold assessment project, the assessment consultant or company shall ensure that all employees who engage in assessment activities and who will be, or are anticipated to be, exposed to mold are provided with, fit tested for, and trained on the appropriate use and care of the specified PPE.

The assessment consultant or company must document successful completion of the training before the employees perform regulated activities. If samples for laboratory analysis are collected during the assessment:

- Sampling must be performed according to nationally accepted methods
- Preservation methods shall be implemented for all samples where necessary
- Proper sample documentation, including the sampling method, the sample identification code, each location and material sampled, the date collected, the name of the person who collected the samples, and the project name or number must be recorded for each sample
- Proper chain of custody procedures must be used
- Samples must be analyzed by a laboratory that is licensed under §295.317 (relating to Mold Analysis Laboratory: Licensing Requirements)

An assessment consultant shall prepare a mold remediation protocol that is specific to each remediation project and provide the protocol to the client before the remediation begins. The mold remediation protocol must specify:

- The rooms or areas where the work will be performed
- The estimated quantities of materials to be cleaned or removed
- The methods to be used for each type of remediation in each type of area
- The PPE to be used by remediators—a minimum of an N95 respirator is recommended during mold-related activities when mold growth could or would be disturbed; using professional judgment, a consultant may specify additional or more protective PPE if he or she determines that it is warranted
- The proposed types of containment, as that term is defined in §295.302(9) (relating to Definitions) and as described in subsection (g), to be used during the project in each type of area
- The proposed clearance procedures and criteria, as described in subsection (i) for each type of remediation in each type of area

A mold assessment consultant shall consider whether to recommend to a client that, before remediation begins, the client should inform building occupants of mold-related activities that will disturb or will have the potential to disturb areas of mold contamination.

Containment must be specified in a mold remediation protocol when the mold contamination affects a total surface area of 25 contiguous square feet or more for the project. Containment is not required if only persons who are licensed or registered under this subchapter occupy the building in which the remediation takes place at any time between the start date and stop date for the project as specified on the notification required under §295.325 (relating to Notifications). The containment specified in the remediation protocol must prevent the spread of mold to areas of the building outside the containment under normal conditions of use. If walk-in containment is used, supply-and-return air vents must be blocked, and air pressure within the walk-in containment must be lower than the pressure in building areas adjacent to the containment.

An assessment consultant who indicates in a remediation protocol that a disinfectant, biocide, or antimicrobial coating will be used on a mold remediation project shall indicate a specific product or brand only if it is registered by the U.S. Environmental Protection Agency (EPA) for the intended use and if the use is consistent with the manufacturer's labeling instructions. A decision by an assessment consultant to use such products must take into account the potential for occupant sensitivities and possible adverse reactions to chemicals that have the potential to be off-gassed from surfaces coated with such products.

In the remediation protocol for the project, the assessment consultant shall specify:

- At least one nationally recognized analytical method for use within each remediated area to determine whether the mold contamination identified for the project has been remediated as outlined in the remediation protocol
- The criteria to be used for evaluating analytical results to determine whether the remediation project passes clearance
- That post-remediation assessment shall be conducted while walk-in containment is in place, if walk-in containment is specified for the project

- The procedures to be used in determining whether the underlying cause of the mold identified for the project has been remediated so that it is reasonably certain that the mold will not return from that same cause

Mold Remediation

A remediation contractor shall prepare a mold remediation work plan that is specific to each project, fulfills all the requirements of the mold remediation protocol, and provides specific instructions and/or standard operating procedures for how a mold remediation project will be performed. The remediation contractor shall provide the mold remediation work plan to the client before site preparation work begins.

If an assessment consultant specifies in the mold remediation protocol that PPE is required for the project, the remediation contractor or company shall provide the specified PPE to all employees who engage in remediation activities and who will, or are anticipated to, disturb or remove mold contamination, when the mold affects a total surface area for the project of 25 contiguous feet or more. The recommended minimum PPE is an N95 respirator. Each employee who is provided PPE must receive training on the appropriate use and care of the provided PPE. The remediation contractor or company must document successful completion of the training before the employee performs regulated activities.

The containment specified in the remediation protocol must be used on a mold remediation project when the mold affects a total surface area of 25 contiguous square feet or more for the project. Containment is not required if only persons who are licensed or registered under this subchapter occupy the building in which the remediation takes place at any time between the start date and stop date for the project as specified on the notification required under §295.325 (relating to Notifications). The containment, when constructed as described in the remediation work plan and under normal conditions of use, must prevent the spread of mold to areas outside the containment. If walk-in containment is used, supply-and-return air vents must be blocked, and air pressure within the walk-in containment must be lower than the pressure in building areas adjacent to the containment.

Signs advising that a mold remediation project is in progress shall be displayed at all accessible entrances to remediation areas. The signs shall be at least 8 × 10 inches in size and shall bear the words "NOTICE: Mold Remediation Project in Progress" in black on a yellow background. The text of the signs must be legible from a distance of 10 feet.

No person shall remove or dismantle any walk-in containment structures or materials from a project site prior to receipt by the licensed mold remediation contractor or remediation company overseeing the project of a written notice from a licensed mold assessment consultant that the project has achieved clearance as described under §295.324 (relating to Post-Remediation Assessment and Clearance).

Disinfectants, biocides, and antimicrobial coatings may be used only if their use is specified in a mold remediation protocol, if they are registered by the EPA for the intended use, and if the use is consistent with the manufacturer's labeling instructions. If a protocol specifies the use of such a product but does not specify the brand or type of product, a remediation contractor may select the brand or type of product to be used, subject to the other provisions of this subsection. A decision by an assessment consultant or remediation contractor to use such a product must take into account the potential for occupant sensitivities and possible adverse reactions to chemicals that have the potential to be off-gassed from surfaces coated with the product. A person who applies a biocide to wood to control a wood-infesting organism must be licensed by the Texas Structural Pest Control Board as provided under the Texas Occupations Code, Chapter 1951 (relating to Structural Pest Control) unless exempt under the Texas Occupations Code, Chapter 1951, Subchapter B (relating to Exemptions).

Remediation of HVAC Systems

All provisions of §295.321 (relating to Minimum Work Practices and Procedures for Mold Assessment) shall apply to the assessment of mold in HVAC systems. All provisions of §295.322 (relating to Minimum Work Practices and Procedures for Mold Remediation) shall apply to the remediation of mold in HVAC systems.

A licensee or registered worker under this subchapter may apply a disinfectant, biocide, or antimicrobial coating in an HVAC system only if its use is specified in a mold remediation protocol, if it is registered by EPA for the intended use, and if the use is consistent with the manufacturer's labeling instructions. The licensee or registered worker shall apply the product only after the building owner or manager has been provided a material safety data sheet for the product, has agreed to the application, and has notified building occupants in potentially affected areas prior to the application. The licensee or registered worker shall follow all manufacturers's label directions when using the product.

Persons who perform air conditioning and refrigeration contracting (including the repair, maintenance, service, or modification of equipment or a product in an environmental air conditioning system, a commercial refrigeration system, or a process cooling or heating system) must be licensed by the Texas Department of Licensing and Registration, as provided under the Texas Occupations Code, Chapter 1302 (relating to Air Conditioning and Refrigeration Contractors). A person who performs biomedical remediation as defined under the Revenue and Tax Code, 16 TAC, §75.10(5) (relating to Definitions), must be licensed by the Texas Department of Licensing and Regulation in accordance with 16 TAC, Chapter 75 (relating to Air Conditioning and Refrigeration Contractor License Law) unless exempt under 16 TAC, §75.30 (relating to Exemptions) or 16 TAC, §75.100 (relating to Technical Requirements).

Post-Remediation Assessment and Clearance

For a remediation project to achieve clearance, a licensed mold assessment consultant shall conduct a post-remediation assessment using visual, procedural, and analytical methods. If walk-in containment is used during remediation, the post-remediation assessment shall be conducted while the walk-in containment is in place. The post-remediation assessment shall determine whether the work area is free from all visible mold and wood rot; and all work has been completed in compliance with the remediation protocol and remediation work plan and meets clearance criteria specified in the protocol.

Post-remediation assessment shall, to the extent feasible, determine that the underlying cause of the mold has been remediated so that it is reasonably certain that the mold will not return from that remediated cause. The assessment consultant shall perform a visual, procedural, and analytical evaluation in each remediated area to determine whether the mold contamination identified for the project has been remediated as outlined in the remediation protocol. The consultant shall use only the analytical methods and the criteria for evaluating analytical results that were specified in the remediation protocol, unless circumstances beyond the control of the consultant and the remediation contractor or company necessitate alternative analytical methods or criteria. The consultant shall provide to the client written documentation of the need for any deviation from the remediation protocol and the alternative analytical methods and criteria selected, and shall obtain approval from the client for their use, before proceeding with the post-remediation assessment.

Where visual inspection reveals deficiencies sufficient to fail clearance, analytical methods need not be used. An assessment consultant who determines that remediation has been successful shall issue a written passed clearance report to the client at the conclusion of each mold remediation project. The report must include the following:

- A description of relevant worksite observations
- The type and location of all measurements made and samples collected at the worksite
- All data obtained at the worksite, including temperature, humidity, and material moisture readings
- The results of analytical evaluation of the samples collected at the worksite
- Copies of all photographs the consultant took
- A clear statement that the project has passed clearance

If the mold assessment consultant determines that remediation has not been successful and ceases to be involved with the project before the project passes clearance, the consultant shall issue a written final status report to the client and to the remediation contractor or company performing the project. The status report must include the items listed in subsections (d)(1) through (5) of this section and any conclusions that the consultant has drawn.

▶ NOTIFICATIONS

A mold remediation contractor or company shall notify the department of a mold remediation, as defined in §295.302(28) (relating to Definitions), when mold contamination affects a total surface area of 25 contiguous square feet or more. Notification shall be received by the Department of State Health Services, Environmental Health Notifications Group no less than five working days (not calendar days) prior to the anticipated start date of the mold remediation and shall be submitted by United States Postal Service, commercial delivery service, hand-delivery, electronic mail (email), or facsimile on a form specified by the department and available on its website. The form must be filled out completely and properly. Blanks that do not apply shall be marked "N/A." The N/A designation will not be accepted for identification of the work site, building description, building owner, individuals required to be identified on the notification form, start and stop dates, or scheduled hours of mold remediation. A signature of the responsible person is required on each notification form. The contractor or company shall retain a confirmation that the department received the notification.

When mold remediation activity is rescheduled to start later than the date or hours contained in the most recent notice, the regional office of the department shall be notified by telephone as soon as possible but prior to the start date on the most recent notice. A written amended notification is required immediately following the telephone notification and shall be emailed, faxed, or overnight mailed to the Environmental Health Notifications Group within the Inspection Unit, Environmental and Consumer Safety Section of the department.

When mold remediation activities begin on a date earlier than the date contained in the notice, the department shall be provided with written notice of the new start date at least five working days before the start of work unless the provisions of subsection (e) of this section apply. The licensee shall confirm with the department by phone that the notice is received five working days before the start of work.

In no event shall mold remediation begin or be completed on a date other than the date contained in the written notice except for operations covered under subsection (e) of this section. Amendments to start-date changes must be submitted as required in subsections (b) and (c) of this section. An amendment is required for any stop dates that change by more than one workday. The contractor or company shall provide schedule changes to the department no less than 24 hours prior to the most recent stop date or the new stop date, whichever comes first. Changes less than five days in advance shall be confirmed with the appropriate department regional office by telephone, facsimile, or email and followed up in writing to the department's central office.

In an emergency, notification to the department shall be made as soon as practicable but not later than the following business day after the license holder identifies the emergency. Initial notification shall be made to the department's central office either immediately by telephone, followed by formal notification on the department's notification form, or immediately by facsimile on the department's notification form. The contractor or company shall retain a confirmation that the notification was received by the department. Emergencies shall be documented. An emergency exists if a delay in mold remediation services in response to a water damage occurrence would increase mold contamination.

For each initial notification of a mold remediation project, the mold remediation contractor or company shall remit to the department a fee of $100, except that the fee shall be $25 for a remediation project in an owner-occupied residential dwelling unit. Amendments to a notification shall not require a separate fee.

The department shall send an invoice for the required fee to the contractor or company after the department has received the notification. Payment must be remitted in the manner instructed on the invoice no later than 60 working days following the date on the notification invoice. Failure to pay the required fee after an invoice has been sent is a violation, and the department may seek administrative penalties as listed in §295.331 (relating to Compliance: Administrative Penalty).

▶ RECORD RETENTION

Records and documents required by this section shall be retained for the time specified in subsection (b)(2) of this section for remediators and subsection (c)(2) of this section for assessors, subsection (d) of this section for mold analysis laboratories, and subsection (e)(1) of this section for training providers. Such records and documents shall be made available for inspection by the department or any law enforcement agency immediately on request. Licensees and accredited training providers who cease to do business shall notify the department in writing 30 days prior to such event to advise how they will maintain all records during the minimum three-year retention period. The department, on receipt of such notification and at its option, may provide instructions for how the records shall be maintained during the required retention period. A licensee or accredited person shall notify the department that it has complied with the department's instructions within 30 days of their receipt or make other arrangements approved by the department. Failure to comply may result in disciplinary action.

A licensed mold remediation company shall maintain the records listed in paragraphs (1) and (2) of this subsection for each mold remediation project performed by the company and the records listed in paragraph (4) of this subsection for each remediation worker training session provided by the company. A licensed mold remediation contractor who is not employed by a company shall personally maintain the records listed in paragraphs (1) and (2) of this subsection for each mold remediation project performed by the contractor and the records listed in paragraph (4) of this subsection for each remediation worker training session provided by the mold remediation contractor. (See Box 8.4.)

A remediation contractor or company may maintain the records required under paragraphs (1) and (2) of this subsection in an electronic format rather than as paper documents. A remediation contractor or company who maintains the required records in an electronic format must provide paper copies of records to a department inspector during an inspection if requested to do so by the inspector.

A licensed mold remediation contractor or remediation company who trains employees to meet the requirements under §295.320(d) (relating to Training: Required Mold Training Courses) shall maintain copies of the required training documents at a central location at its Texas office.

A licensed mold assessment company shall maintain the following records and documents at a central location at its

> **Box 8.4 Records and Documents to Maintain**
>
> A licensed mold remediation contractor shall maintain the records and documents onsite at a project for its duration.
>
> - A current copy of the mold remediation work plan and all mold remediation protocols used in the preparation of the work plan
> - A listing of the names and license/registration numbers of all individuals working on the remediation project
>
> A licensed mold remediation company shall maintain the following records and documents at a central location at its Texas office for three years after the stop date of each project that the company performs. A licensed mold remediation contractor not employed by a company shall maintain the following records and documents at a central location at his or her Texas office for three years following the stop date of each project that the contractor performs:
>
> - A copy of the mold remediation work plan specified under paragraph (1)(A) of this subsection
> - Photographs of the scene of the mold remediation taken before and after the remediation
> - The written contract between the mold remediation company or remediation contractor and the client, and any written contracts related to the mold remediation project between the company or contractor and any other party
> - All invoices issued regarding the mold remediation
> - Copies of all certificates of mold remediation issued by the company or contractor

Texas office for the time period required under paragraph (2) of this subsection for each project that the company performs. A licensed mold assessment consultant not employed by a company shall maintain the following records and documents at a central location at his or her Texas office for the time period required under paragraph (2) of this subsection for each project that the contractor performs:

- The name and mold credential number of each of its employees who worked on the project and a description of each employee's involvement with the project
- The written contract between the mold assessment company or consultant and the client
- All invoices issued regarding the mold assessment
- Copies of all laboratory reports and sample analyses
- Copies of all photographs required under §295.324 (relating to Post-Remediation Assessment and Clearance)

- Copies of all mold remediation protocols and changes prepared as a result of mold assessment activities
- Copies of all passed clearance reports issued by the company or consultant

For each project, a licensed mold assessment company or consultant shall maintain all the records listed in paragraph (1) of this subsection until:

- The company or consultant issues a mold assessment report, management plan, or remediation protocol to a client, if the company or consultant performs only the initial assessment for the project
- The company or consultant issues the final status report to the client, if a final status report is issued
- The company or consultant provides the signed Certificate of Mold Damage Remediation to a mold remediation contractor or company, if a certificate is provided

A licensed mold analysis laboratory shall maintain copies of the results, including the sample identification number, of all analyses performed as part of a mold assessment or mold remediation for three years from the date of the samples' analysis.

Accredited training providers shall comply with the following record-keeping requirements. The training provider shall maintain the records in a manner that allows verification of the required information by the department. The training provider shall maintain records for at least three years from the date of the class in accordance with §295.318(f)(8) and (9) (relating to Mold Training Provider: Accreditation).

A training provider may maintain the records required under paragraph (1) of this subsection in an electronic format rather than as paper documents. A training provider who maintains the required records in an electronic format must provide paper copies of records to a department inspector during an inspection if requested to do so by the inspector.

▶ DOCUMENTING MOLD REMEDIATION

Not later than one week after completion of a mold remediation project, the licensed remediation contractor or company shall provide the property owner with copies of required photographs of the scene of the mold remediation taken before and after the remediation.

Not later than the 10th day after the project stop date, the licensed mold remediation contractor or company shall provide a Certificate of Mold Damage Remediation to the property owner on a form adopted by the Texas Commissioner of Insurance. The certificate must include the following:

- A statement by a licensed mold assessment consultant (not the licensed mold remediator) that based on visual, procedural, and analytical evaluation, the mold contamination identified for the project has been remediated as outlined in the mold remediation protocol
- A statement on the certificate that the underlying cause of the mold has been remediated, if the licensed mold assessment consultant determines that the underlying cause of the mold has been remediated so that it is reasonably certain that the mold will not return from that same cause
- Copies of the completed certificate shall be retained in the business files of the assessment consultant/company and the remediation contractor/company

If a property owner sells the property, the property owner shall provide to the buyer a copy of each Certificate of Mold Damage Remediation issued for the property under this section during the five years preceding the date the property owner sells the property.

▶ **COMPLIANCE**

The department may inspect or investigate the business practices of any persons involved with mold-related activity for compliance with this subchapter.

Inspections and Investigations

A department representative, on presenting a department identification card, shall have the right to enter at all reasonable times any area or environment—including but not limited to any containment area, building, construction site, storage or office area, or vehicle—to review records; to question any person; or to locate, identify, or assess areas of mold growth for the purpose of inspection and investigation for compliance with this subchapter.

A department representative conducting official duties is not required to notify in advance or seek permission to conduct inspections or investigations. It is a violation for any person to interfere with, deny, or delay an inspection or investigation conducted by a department representative. A department representative shall not be impeded or refused entry in the course of official duties by reason of any regulatory or contractual specification.

Reprimand, Suspension, Revocation, Probation

After notice of the opportunity for a hearing in accordance with subsection (d) of this section, the department may take any of the disciplinary actions outlined in subsection (c) of this section. A person who is denied a credential for failure to meet the qualifications under this subchapter is ineligible to reapply until such time as all qualifications are met. A suspension shall be for a period of not more than two years. A person whose application or credential has been revoked shall be ineligible to reapply for any mold-related credential for up to three years.

The department may issue an administrative penalty as described in §295.331 (relating to Compliance: Administrative Penalty), deny an application, suspend, suspend with probationary terms, or revoke a credential of a person who:

- Fails to comply with this subchapter
- Has fraudulently or deceptively obtained or attempted to obtain a credential, ID card, or approval, including engaging in misconduct or dishonesty during the state licensing examination, such as cheating or having another person take or attempt to take the examination for that person
- Duplicates or allows another person to duplicate a credential, ID card, or approval
- Uses a credential issued to another person or allows any other person to use a credential, ID card, or approval not issued to that other person
- Falsifies records for mold-related activities that the department requires the person to create, submit, or maintain
- Is convicted of a felony or misdemeanor arising from mold-related activity

The contested case hearing provisions of the Administrative Procedure Act (Texas Government Code, Chapter 2001) and the formal hearing procedures of the department shall apply to any enforcement action under this section. A person charged with a violation shall be notified of the alleged violation, the grounds on which any disciplinary action is based, the proposed penalty, and the opportunity to request a hearing.

Administrative Penalty

If a person violates the Administrative Procedure Act, this subchapter or an order, the department may assess an administrative penalty. The penalty shall not exceed $5000 per violation except as indicated. Each day a violation continues will be considered a separate violation for violations listed in subsections (d)(1)(A) and (B) and (d)(2)(A) and (B) of this section. The department may reduce or enhance penalties as warranted.

In assessing administrative penalties, including reductions or enhancements, the department shall consider:

- Whether the violation was committed knowingly, intentionally, or fraudulently
- The seriousness of the violation
- Any hazard created to the public health and safety
- The person's history of previous violations
- Any other matter that justice may require, including demonstrated good faith

Violations shall be placed in one of three severity levels.

Severity Level I violations have or may have a direct negative impact on public health or welfare. This category includes fraud and misrepresentation. The penalty for a Level I violation may be up to $5000 per violation. Violations listed in subparagraphs (A) and (B) of this paragraph may be assessed at up to $5000 per violation per day. Examples include but are not limited to:

- Working without a valid credential, ID card, or approval or with a credential or ID card that has been expired for more than one month
- Engaging in a conflict of interest as described in §295.307(a)(1) and (2) (relating to Conflict of Interest)
- Engaging in misconduct or dishonesty during the state licensing examination
- Submitting a forged or altered training certificate

- Offering training required under this subchapter without valid department approval of the course, the instructor(s), or the guest speaker(s), except as provided under §295.320(d)(1)(A) of this subchapter (relating to Training: Required Mold Training Courses)
- Providing training certificates for a course required by the department to persons who have not successfully completed the course
- Failing to meet the insurance requirements of §295.309 (relating to Licensing: Insurance Requirements)
- Failure of an assessment consultant to specify containment in a mold remediation protocol
- Failure of a remediator to use the containment specified in the mold remediation protocol for the project

Severity Level II violations could compromise public health or welfare. The maximum penalty for Level II violations is $2500 per violation. Violations listed in subparagraphs (A) and (B) of this paragraph may be assessed at up to $2500 per violation per day. Examples include but are not limited to:

- Working with a credential or ID card that has been expired for one month or less
- Failing to disclose an ownership interest as required in §295.307(b)
- Failing to submit a timely notification
- Failure to conduct a training course as specified under §295.320 (relating to Training: Required Mold Training Courses)
- Failure of a credentialed person to maintain current required training

Severity Level III violations, while not having a direct negative impact on health or welfare, could lead to more serious circumstances. The maximum penalty for Level III violations is $1000 per violation. Examples include but are not limited to:

- Failure to provide the department Consumer Mold Information Sheet as required under §295.306 (relating to Credentials: General Responsibilities)
- Failure to have a department-issued ID card at a jobsite
- Submitting an incorrect or improper notification
- Failure of a training provider to submit information to the department regarding training course schedules or to notify the department of cancellations within the specified time periods

- Failure of a training provider to submit course completion information within the time period specified in §295.319(f)(7) (relating to Mold Training Provider: Accreditation)
- Failure of a remediation company, remediation contractor, or training provider to submit worker training information within the time period specified in §295.320(d) (relating to Mold Training Provider: Accreditation)
- Failure of a training provider to maintain the required trainee-instructor ratio in a training course

The commissioner may choose not to impose an administrative penalty under §295.331 (relating to Compliance: Administrative Penalty) if, not later than the 10th day after the date on a written notice of a violation as provided under §295.333 (relating to Compliance: Notice, Opportunity for Hearing, Order), the person charged with the violation provides evidence satisfactory to the department that the circumstances giving rise to the violation have been corrected and all actual damages are paid.

This section does not apply to a violation alleged under the Texas Occupations Code, Chapter 1958, §1958.101 (relating to License Required; Rules); §295.305(a) and (b) (relating to Credentials: General Conditions); the Texas Occupations Code, Chapter 1958, §1958.155 (relating to Conflict of Interest; Disclosure Required); or §295.307 (relating to Conflict of Interest and Disclosure Requirement).

Notice, Opportunity for Hearing, Order

The commissioner shall impose an administrative penalty under this subchapter only after a person is given written notice of the opportunity for a hearing conducted in accordance with the Administrative Procedure Act (Texas Government Code, Chapter 2001) and the department's formal hearing procedures.

The written notice of violation must state the facts that constitute the alleged violation, the law or rule that has been violated, the proposed penalty, and the opportunity for a hearing.

If a hearing is held, the commissioner shall make findings of fact and issue a written decision as to the occurrence of the violation and the amount of any penalty that is warranted.

If a person fails to request a hearing, the commissioner, after determining that a violation occurred and the amount of penalty warranted, is authorized to impose a penalty and issue an order requiring the person to pay the penalty imposed. Not later than the 30th day after the date the commissioner issues an order, the commissioner shall inform the person of the amount of any penalty imposed.

The commissioner is authorized to consolidate a hearing under this section with another proceeding. Not later than the 30th day after the date the commissioner's decision or order concerning an administrative penalty assessed under §295.331 (relating to Compliance: Administrative Penalty) becomes final as provided by the Texas Government Code, §2001.144, (relating to Decisions: When Final) to the person against whom the penalty is assessed, that person either shall pay the administrative penalty or shall file a petition for judicial review.

A person who files a petition for judicial review can stay the enforcement of the penalty either by paying the penalty to the commissioner for placement in an escrow account or by giving the commissioner a bond, in a form approved by the commissioner, for the amount of the penalty that is effective until judicial review of the commissioner's decision or order is final.

Collection of Administrative Penalty, Judicial Review

At the request of the commissioner, the Texas Attorney General is authorized to bring a civil action to recover an administrative penalty imposed under §295.331 (relating to Compliance: Administrative Penalty).

Judicial review of a decision or order of the commissioner imposing a penalty is instituted by filing a petition with a district court in Travis County and is under the substantial evidence rule as provided by the Texas Government Code, Chapter 2001, Subchapter G (relating to Contested Cases: Judicial Review).

After judicial review, if the administrative penalty is reduced or is not upheld by the court, not later than the 30th day after the date of the determination, the commissioner shall do one of the following:

- Remit the appropriate amount, plus accrued interest, to a person who paid the penalty as provided under §295.334 (relating to Compliance: Options Following Administrative Order)
- Execute a release of a bond provided under §295.334(b) to the person who gave the bond

A person who violates the Act or this subchapter is liable for a civil penalty in an amount not to exceed $2000 for the first violation or $10,000 for a second or subsequent violation. The commissioner may request the Texas Attorney General or the district, county, or city attorney having jurisdiction to bring an action to collect a civil penalty under this section.

The commissioner may request the Texas Attorney General or the district, county, or city attorney having jurisdiction to bring an action for a restraining order, injunction, or other relief that the court determines appropriate if it appears to the department that a person is violating or has violated the Act or this subchapter.

A property owner is not liable for damages related to mold remediation on a property if a Certificate of Mold Damage Remediation has been issued under §295.327 (relating to Photographs, Certificate of Mold Damage Remediation, Duty of Property Owner) for that property and the damages accrued on or before the date of the issuance of the Certificate of Mold Damage Remediation.

A person is not liable in a civil lawsuit for damages related to a decision to allow occupancy of a property after mold remediation has been performed on the property if a Certificate of Mold Damage Remediation has been issued under §295.327 for the property, the property is owned or occupied by a governmental entity, including a school, and the decision was made by the owner, the occupier, or any person authorized by the owner or occupier to make the decision.

As you can see, regulations for mold remediation and abatement can be complex. This is not the case everywhere. However, if the rules are in effect in your region, they must be adhered to. The examples given here are based on the requirements of Texas. Different states have varying degrees of regulation, and some do not regulate working with mold at all. It is in your best interest to know what you are required to comply with before you do any work.

Glossary

AHU *See* Air-handling unit.

Air cleaning An IAQ control strategy to remove various airborne particulates and/or gases from the air. The three types of air cleaning most commonly used are particulate filtration, electrostatic precipitation, and gas sorption.

Air exchange rate The rate at which outside air replaces indoor air in a space. Expressed in one of two ways: the number of changes of outside air per unit of time—air changes per hour (ACH); or the rate at which a volume of outside air enters per unit of time—cubic feet per minute (cfm).

Air-handling unit Equipment that includes a blower or fan, heating and/or cooling coils, and related equipment such as controls, condensate drain pans, and air filters. Does not include ductwork, registers or grilles, or boilers and chillers.

Animal dander Tiny scales of animal skin.

Allergen A substance capable of causing an allergic reaction because of an individual's sensitivity to that substance.

Allergic rhinitis Inflammation of the mucous membranes in the nose caused by an allergic reaction.

Antimicrobial Agent that kills microbial growth (i.e., chemical or substance that kills mold or other organisms). *See also* Biocide; Disinfectants; Fungicide; Sanitizer; Sterilizer.

Biological contaminants (1) Living organisms, such as viruses, bacteria, or mold (fungi); (2) the remains of living organisms; or (3) debris from or pieces of dead organisms. Biological contaminants can be small enough to be inhaled, and may cause many types of health effects including allergic reactions and respiratory disorders. Also referred to as "microbiologicals" or "microbials."

Biocide A substance or chemical that kills organisms such as mold.

Breathing zone Area of a room in which occupants breathe as they stand, sit, or lie down.

Building envelope Elements of the building, including all external building materials, windows, and walls, that enclose the internal space.

Building-related illness (BRI)   Diagnosable illness with symptoms that can be identified and where the cause can be directly attributed to airborne building pollutants (e.g., Legionnaire's disease, hypersensitivity pneumonitis). Also: A discrete, identifiable disease or illness that can be traced to a specific pollutant or source within a building. *See also* Sick building syndrome.

Ceiling plenum   Space between a suspended ceiling and the floor above that may have mechanical and electrical equipment in it and that is used as part of the air distribution system. The space is usually designed to be under negative pressure.

Central air-handling unit (central AHU)   This is the same as an air handling unit, but serves more than one area.

CFM   Cubic feet per minute. The amount of air, in cubic feet, that flows through a given space in one minute; 1 cfm equals approximately 2 liters per second (l/s).

Chemical sensitization   Evidence suggests that some people may develop health problems characterized by effects such as dizziness, eye and throat irritation, chest tightness, and nasal congestion that appear whenever they are exposed to certain chemicals. People may react to even trace amounts of chemicals to which they have become "sensitized."

CO   Carbon monoxide.

CO$_2$ Carbon dioxide.

Conditioned air   Air that has been heated, cooled, humidified, or dehumidified to maintain an interior space within the "comfort zone." (Sometimes referred to as "tempered" air.)

Dampers   Controls that vary airflow through an air outlet, inlet, or duct. A damper position may be immovable, manually adjustable, or part of an automated control system.

Diffusers and grilles   Components of the ventilation system that distribute and return air to promote air circulation in the occupied space. As used in this document, supply air enters a space through a diffuser or vent and return air leaves a space through a grille.

Disinfectants   One of three groups of antimicrobials registered by EPA for public health uses. EPA considers an antimicrobial to be a disinfectant when it destroys or irreversibly inactivates infectious or other undesirable organisms, but not necessarily their spores. EPA registers three types of disinfectant products based on submitted efficacy data: limited, general or broad spectrum, and hospital disinfectant.

Environmental agents Conditions other than indoor air contaminants that cause stress, comfort, and/or health problems (e.g., humidity extremes, drafts, lack of air circulation, noise, and overcrowding).

Environmental tobacco smoke (ETS) Mixture of smoke from the burning end of a cigarette, pipe, or cigar and smoke exhaled by the smoker (also secondhand smoke (SHS) or passive smoking).

Exhaust ventilation Mechanical removal of air from a portion of a building (e.g., piece of equipment, room, or general area).

Flow hood Device that easily measures airflow quantity, typically up to 2500 cfm.

Fungi A separate kingdom comprising living things that are neither animals nor plants. The kingdom *Fungi* includes molds, yeasts, mushrooms, and puffballs. In this book, the terms fungi and mold are used interchangeably.

Fungicide A substance or chemical that kills fungi.

Governmental In the case of building codes, these are the state or local organizations/agencies responsible for building code enforcement.

HEPA High-efficiency particulate air (filter).

Humidifier fever A respiratory illness caused by exposure to toxins from microorganisms found in wet or moist areas in humidifiers and air conditioners. Also called air-conditioner or ventilation fever.

HVAC Heating, ventilation, and air-conditioning system.

Hypersensitivity Great or excessive sensitivity.

Hypersensitivity diseases These are characterized by allergic responses to pollutants. The ones most clearly associated with IAQ are asthma, rhinitis, and hypersensitivity pneumonitis.

Hypersensitivity pneumonitis (HP) A group of rare, but serious, respiratory diseases that cause lung inflammation (specifically granulomatous cells). Most forms of hypersensitivity pneumonitis are caused by the inhalation of organic dusts, including molds. HP disease is progressive as long as one is exposed to the causative agent such as mold.

IAQ Indoor air quality is a term referring to the air quality within and around buildings and structures, especially as it relates to the health and comfort of building occupants.

IAQ backgrounder   A component of the *IAQ Tools for Schools* Action Kit that provides a general introduction to IAQ issues, as well as IAQ program implementation information.

IAQ coordinator   An individual at the school and/or school district level who provides leadership and coordination of IAQ activities.

IAQ checklist   A component of the *IAQ Tools for Schools* Action Kit containing information and suggested easy-to-do activities for school staff to improve or maintain good indoor air quality. Each Activity Guide focuses on topic areas and actions that are targeted to particular school staff. The checklists are to be completed by the staff and returned to the IAQ coordinator as a record of activities completed and assistance requested.

IAQ management plan   A component of the *IAQ Tools for Schools* Action Kit, specifically, a set of flexible and specific steps for preventing and resolving IAQ problems.

IAQ team   People who have a direct impact on IAQ in the schools (school staff, administrators, school board members, students, and parents) and who implement the IAQ Action Kits.

IPM   Integrated pest management.

Indicator compounds   Chemical compounds, such as carbon dioxide, whose presence at certain concentrations may be used to estimate certain building conditions (e.g., airflow, presence of sources).

Indoor air pollutant   Particles and dust, fibers, mists, bioaerosols, and gases or vapors.

Indoor air quality   *See* IAQ.

Make-up air   *See* Outdoor air supply.

Map of radon zones   A U.S. EPA publication depicting areas of differing radon potential in both map form and in state specific booklets.

MCS   *See* Multiple chemical sensitivity.

Mechanically ventilated crawlspace system   A system designed to increase ventilation within a crawlspace, achieve higher air pressure in the crawlspace relative to air pressure in the soil beneath the crawlspace, or achieve lower air pressure in the crawlspace relative to air pressure in the living spaces, by use of a fan.

Microbiologicals   *See* Biological contaminants.

Model Code Organizations   Includes the following agencies and the model building codes they promulgate:

• Building Officials and Code Administrators International, Inc. (BOCA National Building Code/1993 and BOCA National Mechanical Code/1993)

- International Conference of Building Officials (Uniform Building Code/1991 and Uniform Mechanical Code/1991)
- Southern Building Code Congress, International, Inc.; that is, Standard Building Code/1991 and Standard Mechanical Code/1991
- Council of American Building Officials (CABO) One- and Two-Family Dwelling Code-1992 and CABO Model Energy Code-1993

Mold   A group of organisms that belong to the kingdom Fungi. In this book, the terms fungi and mold are used interchangeably.

MVOC (microbial volatile organic compound)   A chemical made by mold that is a gas at room temperature and may have a moldy or musty odor.

Mycotoxin A toxin produced by a mold.

Negative pressure   A condition that exists when less air is supplied to a space than is exhausted from the space, so the air pressure within that space is less than that in surrounding areas. Under this condition, if an opening exists, air will flow from surrounding areas into the negatively pressurized space.

Organic compounds   These are chemicals that contain carbon. Volatile organic compounds vaporize at room temperature and pressure. They are found in many indoor sources, including many common household products and building materials.

Outdoor air supply   Air brought into a building from the outdoors (often through the ventilation system) that has not been previously circulated through the system. Also known as "make-up air."

PELs   Permissible exposure limits—standards set by the Occupational, Safety and Health Administration (OSHA).

Plenum   Air compartment connected to a duct or ducts.

PM   *See* Preventive maintenance.

Pollutant pathways   Avenues for distribution of pollutants in a building. HVAC systems are the primary pathways in most buildings; however all building components interact to affect how air movement distributes pollutants. Also a term used in the *IAQ Tools for Schools: IAQ Coordinator's Guide.*

Positive pressure   Condition that exists when more air is supplied to a space than is exhausted, so the air pressure within that space is greater than that in surrounding areas. Under this condition, if an opening exists, air will flow from the positively pressurized space into surrounding areas.

PPM   Parts per million; 1 ppm = 10^{-6} or 0.0001% and 1% = 10,000 ppm.

Pressure, static   In flowing air, the total pressure minus velocity pressure. The portion of the pressure that pushes equally in all directions.

Pressure, total   In flowing air, the sum of the static pressure and the velocity pressure.

Pressure, velocity   In flowing air, the pressure due to the velocity and density of the air.

Preventive maintenance (PM)   Regular and systematic inspection, cleaning, and replacement of worn parts, materials, and systems. Preventive maintenance helps to prevent parts, material, and systems failure by ensuring that parts, materials, and systems are in good working order.

Radon (Rn)/Radon decay products   Radon is a radioactive gas formed in the decay of uranium. The radon decay products (also called radon daughters or progeny) can be breathed into the lung where they continue to release radiation as they further decay.

Reentrainment; Reentry   Situation that occurs when the air being exhausted from a building is immediately brought back into the system through the air intake and other openings in the building envelope.

RELs   Recommended exposure limits—recommendations made by the National Institute for Occupational Safety and Health (NIOSH).

Remediate   Fix.

Sanitizer   One of three groups of antimicrobials registered by EPA for public health uses. EPA considers an anti-microbial to be a sanitizer when it reduces but does not necessarily eliminate all the microorganisms on a treated surface. To be a registered sanitizer, the test results for a product must show a reduction of at least 99.9% in the number of each test microorganism over the parallel control.

Short-circuiting   Situation that occurs when the supply air flows to return or exhaust grilles before entering the breathing zone (area of a room where people are). To avoid short-circuiting, the supply air must be delivered at a temperature and velocity that results in mixing throughout the space.

Sick building syndrome (SBS)   Term that refers to a set of symptoms that affect some number of building occupants during the time they spend in the building and diminish or go away during periods when they leave the building. Cannot be traced to specific pollutants or sources within the building. *See also* Building related illness.

Sources   Sources of indoor air pollutants. Indoor air pollutants can originate within the building or be drawn in from outdoors. Common sources include people, room furnishings such as carpeting, photocopiers, art supplies, etc.

Spore   The means by which molds reproduce. Spores are microscopic. They vary in shape and range from 2 to 100 microns in size. Spores travel in several ways: passively moved by a breeze or water drop, mechanically disturbed (by a person or animal passing by), or actively discharged by the mold (usually under moist conditions or high humidity).

Static pressure   Condition that exists when an equal amount of air is supplied to and exhausted from a space. At static pressure, equilibrium has been reached.

Sterilizer   One of three groups of antimicrobials registered by EPA for public health uses. EPA considers an antimicrobial to be a sterilizer when it destroys or eliminates all forms of bacteria, fungi, viruses, and their spores. Because spores are considered the most difficult form of a microorganism to destroy, EPA considers the term sporicide to be synonymous with "sterilizer."

Sub-membrane depressurization system   A system designed to achieve lower sub-membrane air pressure relative to crawl-space air pressure by use of a fan-powered vent drawing air from under the soil-gas-retarder membrane.

Toxigenic   Producing toxic substances.

TLVs   Threshold limit values—guidelines recommended by the American Conference of Governmental Industrial Hygienists.

TVOCs   Total volatile organic compounds. *See* Volatile organic compounds (VOCs).

Unit ventilator   A fan-coil unit package device for applications in which the use of outdoor- and return-air mixing is intended to satisfy tempering requirements and ventilation needs.

Variable air volume system (VAV)   Air handling system that conditions the air to constant temperature and varies the outside airflow to ensure thermal comfort.

Ventilation air   Defined as the total air, which is a combination of the air brought inside from outdoors and the air that is being recirculated within the building. Sometimes, however, used in reference only to the air brought into the system from outdoors; this book defines this air as "outdoor air ventilation."

Volatile organic compounds (VOCs) Compounds that vaporize (become a gas) at room temperature. Common sources that may emit VOCs into indoor air include housekeeping and maintenance products, and building and furnishing materials. In sufficient quantities, VOCs can cause eye, nose, and throat irritations; headaches, dizziness, visual disorders, or memory impairment. Some are known to cause cancer in animals; some are suspected of causing, or are known to cause, cancer in humans. At present, not much is known about what health effects occur at the levels of VOCs typically found in public and commercial buildings.

Zone The occupied space or group of spaces within a building that has its heating or cooling controlled by a single thermostat.

Resources

Mold Publications

You can order publications about indoor air quality (IAQ) from EPA's National Service Center for Environmental Publications (NSCEP). NSCEP operates a toll-free phone service for publication assistance with live customer service representatives, Monday through Friday from 9:00 A.M. to 5:30 P.M. Eastern time, voicemail after operating hours, or you can fax or email publication requests. Use the EPA Document Number, which is usually bolded or highlighted, when ordering from NSCEP.

A Brief Guide to Mold, Moisture, and Your Home is available in HTML and PDF (20 pp., 278 KB); *www.epa.gov/mold/moldguide.html*

Mold Remediation in Schools and Commercial Buildings is available in HTML and PDF (54 pp., 5 MB); *www.epa.gov/mold/moldguide.html*

U.S. Environmental Protection Agency, National Service Center for Environmental Publications (NSCEP), P.O. Box 42419, Cincinnati, OH 45242-0419; *www.epa.gov/nscep*; phone: 1-800-490-9198; fax: 301-604-3408; email: nscep@bps-lmit.com

Indoor Air Quality Publications and Resources

An Office Building Occupants Guide to Indoor Air Quality: *www.epa.gov/iaq/pubs/occupgd.html*

Biological Pollutants: *www.epa.gov/iaq/biologic.html*

IAQ Building Education and Assessment Model (I-BEAM) updates and expands EPA's existing Building Air Quality guidance and is designed to be a comprehensive state-of-the-art guidance for managing IAQ in commercial buildings. This guidance was designed to be used by building professionals and others interested in IAQ in commercial buildings. I-BEAM contains text, animation/visual, and interactive/calculation components that can be used to perform a number of diverse tasks: *www.epa.gov/iaq/largebldgs/i-beam*

Building Air Quality: A Guide for Building Owners and Facility Managers (BAQ Guide): *www.epa.gov/iaq/largebldgs/baqtoc.html*

Flood Cleanup and the Air in Your Home: *www.epa.gov/iaq/flood*

For more subject-specific links, go to *epa.gov/iaq/schools/links. html, epa.gov/iaq/asthma/links.html, epa.gov/iaq/moreinfo.html,* or *epa.gov/radon/rnlinks.html.*

Search frequently asked questions (FAQs) or submit question and/or comments to the Frequent Questions database. In addition to questions and answers about mold, you can use this database to find information on other topic areas (e.g., asthma, radon, IAQ tools for schools, smoke-free homes, IAQ design tools for schools, and general IAQ issues). You can also use this database to subscribe to new or updated information relating to mold that may be posted on EPA's website.

Antimicrobial Information Hotline: *www.epa.gov/oppad001/*; phone: 703-308-0127; fax: 703-308-6467; Monday–Friday 8:00 A.M.–5:00 P.M. EST; email: Info_Antimicrobial@epa.gov

- The Antimicrobials Information Hotline provides answers to questions concerning current antimicrobial issues (disinfectants, fungicides, others) regulated by the pesticide law, rules, and regulations. These cover interpretation laws, rules, and regulations, and registration and reregistration of antimicrobial chemicals and products. The hotline also provides information on health and safety issues concerning registered antimicrobial products, product labels, and the proper and safe use of these antimicrobial products.

Related Links

The following list of resources includes information created and maintained by other public and private organizations. EPA does not control or guarantee the accuracy, relevance, timeliness, or completeness of this information. The inclusion of such resources is not intended to endorse any views expressed or products or services offered by the reference's author or the organization operating the service on which the reference is maintained.

Asthma and Allergic Diseases

Allergy and Asthma Network Mothers of Asthmatics (AANMA); 800-878-4403/703-641-9595; *www.aanma.org*
- Information on allergies and asthma

American Academy of Allergy, Asthma & Immunology (AAAAI); 800-822-2762; *www.aaaai.org*
- Physician referral directory, information on allergies and asthma

American Lung Association (ALA); 800-LUNG-USA/800-586-4872; *www.lungusa.org*
- Information on allergies and asthma

Asthma and Allergy Foundation of American (AAFA); 800-7-ASTHMA/800-727-8462; *www.aafa.org*
- Information on allergies and asthma

National Institute of Allergy and Infectious Diseases (NIAID); 301-496-5717; *www.niaid.nih.gov*
- Information on allergies and asthma

National Jewish Health; 800-423-8891; *www.nationaljewish.org*
- Information on allergies and asthma

Floods/Flooding

American Red Cross, National Headquarters, 2025 E Street NW, Washington, DC 20006; 202-303-4498/800-REDCROSS/800-733-2767. Repairing Your Flooded Home: *www.redcross.org/www-files/Documents/pdf/Preparedness/file_cont333_lang0_150.pdf*

Federal Emergency Management Agency (FEMA); 800-480-2520; *www.fema.gov*; Flood information: *www.fema.gov/hazard/flood/index.shtm*

University of Minnesota, Department of Environmental Health and Safety; 612-626-5804; Flood information: *www.dehs.umn.edu/iaq.htm*

University of Wisconsin-Extension, The Disaster Network; 608-262-3980; *www.uwex.edu/ces/news/handbook.html*

Alphabetical Listing of Other Links

American Coatings Association, 1500 Rhode Island Ave., NW, Washington, DC 20005; phone: 202-462-6272; fax: 202-462-8549; *aca@paint.org*
- How-To Brochures: Preventing Moisture Damage

American College of Occupational and Environmental Medicine (ACOEM); 847-818-1800; *www.acoem.org*
- Referrals to physicians who have experience with environmental exposures (a members only service)

American Conference of Governmental Industrial Hygienists, Inc. (ACGIH); 513-742-2020; www.acgih.org
- Occupational and environmental health and safety information

American Industrial Hygiene Association (AIHA); 703-849-8888; *www.aiha.org*

American Society of Heating, Refrigerating and Air-Conditioning Engineers, Inc. (ASHRAE); 800-527-4723; *www.ashrae.org*
- Information on engineering issues and indoor air quality

Association of Occupational and Environmental Clinics (AOEC); 202-347-4976; *www.aoec.org*
- Referrals to clinics with physicians who have experience with environmental exposures, including exposure to mold; maintains a database of occupational and environmental cases

Building Ecology, attn: Hal Levin, 2548 Empire Grade, Santa Cruz, CA 95060; phone: 831-425-3946; fax: 831-426-6522; email: hal.levin@buildingecology.com; *www.buildingecology.com/*
- Includes information on mold and dampness with a focus on building ecology

Canada Mortgage and Housing Corporation (CMHC); 613-748-2000 [International]; *www.cmhc-schl.gc.ca/en/*
- Several documents on mold-related topics available including "Fighting Mold—The Homeowner's Guide"; *www.cmhc-schl.gc.ca/en/co/maho/yohoyohe/momo/momo_005.cfm*

Carpet and Rug Institute (CRI); 800-882-8846; *www.carpet-rug.com*
- Carpet maintenance, restoration guidelines for water-damaged carpet, other carpet-related issues

Centers for Disease Control and Prevention (CDC); 800-232-4636; *www.cdc.gov*; CDC's National Center for Environmental Health (NCEH); *www.cdc.gov/nceh*
- Information on health-related topics including asthma, molds in the environment, and occupational health
- "Molds in the Environment" factsheet: *www.cdc.gov/mold/faqs.htm*
- Stachybotrys or *Stachybotrys atra (chartarum)* and health effects: *www.cdc.gov/mold/stachy.htm*

Cornell University Department of Environmental Health and Safety, Material Safety Data Sheets (MSDSs); 607-255-8200 *http://sp.ehs.cornell.edu/lab-research-safety/research-safety/msds/Pages/default.aspx*;
- MSDSs contain information on chemicals or compounds including topics such as health effects, first aid, and protective equipment for people who work with or handle these chemicals. The ~250,000 MSDS files contained in this database are derived from the U.S. Department of Defense MSDS database.
- MSDS sheets maintained by Environmental Health and Safety and other Cornell departments

Energy and Environmental Building Alliance; 952-881-1098; *www.eeba.org*
- Information on energy-efficient and environmentally responsible buildings, humidity/moisture control/vapor barriers

Health Canada, Health Protection Branch, Laboratory Centre for Disease Control, Office of Biosafety; phone: 613-957-2991; *www.hc-sc.gc.ca/ewh-semt/index_e.html*
- Material Safety Data Sheets with health and safety information on infectious microorganisms, including Aspergillus and other molds and airborne biologicals

Institute of Inspection, Cleaning and Restoration Certification (IICRC); 360-693-5675; *www.iicrc.org*
- Information on and standards for the inspection, cleaning, and restoration industry

International Sanitary Supply Association (ISSA); 800-225-4772; *www.issa.com*
- Education and training on cleaning and maintenance

International Society of Cleaning Technicians (ISCT); 800-604-0468; www.isct.com
- Information on cleaning, such as a stain removal guide for carpets

Medical College of Wisconsin, Healthlink, Office of Clinical Informatics, 9200 West Wisconsin Ave, Suite 2975, Milwaukee, WI 53226; phone: 414-955-8296; fax: 414-805-7967; email: healthlink@mcw.edu; "Molds in the (Indoor) Environment": *http://healthlink.mcw.edu/article/1031002357.html*

Mid-Atlantic Environmental Hygiene Resource Center (MEHRC), University City Science Center, 3701 Market Street, 1st Floor, Philadelphia, PA 19104; phone: 215-966-6191; fax: 215-387-6321; *www.mehrc.org*
- Indoor environmental quality training on topics such as mold remediation

National Air Duct Cleaners Association (NADCA); 202-737-2926; *www.nadca.com*
- Duct cleaning information

National Association of the Remodeling Industry (NARI); 847-298-9200; *www.nari.org*
- Consumer information on remodeling, including help finding a professional remodeling contractor

National Environmental Health Association (NEHA), 720 S. Colorado Blvd., Suite 1000-N, Denver, CO 80246–1926; phone: 303-756-9090; fax: 303-691-9490; email: staff@neha.org; *www.neha.org*
- NEHA offers a variety of programs. Expanding beyond its original credential, today the association has seven national credential programs

National Institute of Building Sciences (NIBS); 202-289-7800; *www.nibs.org*
- Information on building regulations, science, and technology

National Institute for Occupational Safety and Health (NIOSH); 800-CDC-INFO/800-232-4636; *www.cdc.gov/niosh*
- Health and safety information with a workplace orientation

National Pesticide Information Center (NPIC); 800-858-7378; *www.npic.orst.edu*
- Information on pesticides/antimicrobial chemicals, including safety and disposal information

The New York City Department of Health and Mental Hygiene, Bureau of Environmental and Occupational Disease Epidemiology; 212-NEW-YORK; Guidelines on Assessment and Remediation of Fungi in Indoor Environments: *www.nyc.gov/html/doh/html/epi/moldrpt1.shtml*
- This document revises and expands the original guidelines to include all fungi (mold). It is based both on a review of the literature regarding fungi and on comments obtained by a review panel consisting of experts in the fields of microbiology and health sciences. It is intended for use by building engineers and management, but is available for general distribution to anyone concerned about fungal contamination, such as environmental consultants, health professionals, or the general public.

Occupational Safety & Health Administration (OSHA); 800-321-OSHA/800-321-6742; *www.osha.gov*; OSHA Mold page: *www.osha.gov/SLTC/molds/*
- Information on worker safety, including topics such as respirator use and safety in the workplace: "A Brief Guide to Mold in the Workplace," *www.osha.gov/dts/shib/shib101003.html*; SHIB 03-10-10

The Restoration Industry Association (RIA), Rockville, MD 20852; 301-231-6505; *www.restorationindustry.org*
- Disaster recovery, water and fire damage, emergency tips, referrals to professionals

Sheet Metal & Air Conditioning Contractors' National Association (SMACNA); 703-803-2980; *www.smacna.org*
- Technical information on topics such as air conditioning and air ducts

Smithsonian Museum Conservation Institute (MCI), Suitland, MD 20746; 301-238-1240; *http://si.edu/mci/*

- Guidelines for caring for and preserving furniture and wooden objects, paper-based materials; preservation studies

University of Minnesota, Department of Environmental Health and Safety; 612-626-5804; *www.dehs.umn.edu/iaq.htm*

- "Fungi in Buildings" (includes "The Fungal Glossary"): *www.dehs.umn.edu/iaq_fib.htm*
- "Fungal Abatement Safe Operating Instruction": *www.dehs.umn.edu/iaq_fasop.htm*

The University of Tulsa, Indoor Air Program, Tulsa, OK 74104; 918-631-5246; *www.utulsa.edu/iaqprogram*

- Courses, classes, and continuing education on indoor air quality

U.S. Department of Housing and Urban Development (HUD), Office of Native American Programs; 202-708-1112; *www.hud.gov/offices/pih/ih/codetalk/onap/*

- Mold and Moisture Prevention: A Guide for Residents in Indian Country: *www.hud.gov/offices/pih/ih/codetalk/docs/moldprevention.pdf*

Western Wood Products Association, Portland, OR 97204; 503-224-3930; *www2.wwpa.org/*

- A trade association representing softwood lumber manufacturers in the 12 western states.
- "Mold and Wood Products" *www.wwpa.org/index_lumberandmold.htm*

Index

Note: Page numbers followed by *b* indicate boxes, *f* indicate figures, *t* indicate tables, and *ge* indicate glossary terms.